Sustainability Assessment

Sustainability Assessment
Context of Resource and Environmental Policy

Mohammad Ali, PhD

Department of Environmental Science and Management
North South University, Dhaka, Bangladesh

AMSTERDAM • BOSTON • HEIDELBERG • LONDON
NEW YORK • OXFORD • PARIS • SAN DIEGO
SAN FRANCISCO • SINGAPORE • SYDNEY • TOKYO
Academic Press is an imprint of Elsevier

Academic Press is an imprint of Elsevier
The Boulevard, Langford Lane, Kidlington, Oxford, OX5 1GB, UK
225 Wyman Street, Waltham, MA 02451, USA

First published 2013

British Library Cataloguing-in-Publication Data
A catalogue record for this book is available from the British Library

Library of Congress Cataloging-in-Publication Data
A catalog record for this book is available from the Library of Congress

ISBN: 978-0-12-407196-4

For information on all Academic Press publications
visit our website at store.elsevier.com

This book has been manufactured using Print On Demand technology. Each copy is produced to order and is limited to black ink. The online version of this book will show color figures where appropriate.

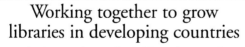

Working together to grow
libraries in developing countries

www.elsevier.com | www.bookaid.org | www.sabre.org

ELSEVIER   BOOK AID
            International   Sabre Foundation

Transferred to Digital Printing in 2012

CONTENTS

LIST OF ABBREVIATIONS

ADB	Asian Development Bank
CCF	Chief Conservator of Forests
CGE	Computable General Equilibrium (A Model)
CIFOR	Center for International Forestry Research
CITES	Convention on International Trade in Endangered Species
EAR	Effectiveness of Activities on Resources (Indicator)
EF	Ecological Footprints
ENR	Energy
ENRINTY	(ENR per Cap)/(GNP per Cap)
ER	Ecological Rucksacks
ESI	Environmental Sustainability Index
FAO	Food and Agricultural Organization
GATT	General Agreement on Tariff and Trade
GDP	Gross Domestic Product
GIS	Geographic Information System
GNP	Gross Net Product
GOB	Government of Bangladesh
IMF	International Monetary Fund
IPAT	Impact, Population, Affluence, Technology (A Model)
IRR	Internal Rate of Return
MoFEC	Ministry of Forests and Estate Crops (Indonesia)
NATO	Nodality, Authority, Treasury and Organization (Government Function)
NCEDR	National Center for Environmental Decision-making Research (USA)
NFR	Need For Resource (Indicator)
NGO	Nongovernmental Organization
NORAD	Norwegian Assistance for Development
NRDA	Natural Resource Dependent Area
NTFP	Nontimber Forest Products
OECD	Organization for Economic Cooperation and Development
PERTs	Political Economic Resource Transfers
PPP (3P)	People, Process, and Perspectives
RAAKS	Rapid Appraisal of Agricultural Knowledge System (A Model)
SDSS	Spatial Decision Support System
SCP	Sustainable Consumption and Production

SOR	State of Resource (Indicator)
UN	United Nations
UNCTAD	United Nations Conference on Trade and Development
UNDP	United Nations Development Programme
UNEP	United Nations Environmental Programme
WB	World Bank
WCED	World Commission on Environment and Development
WRI	World Resources Institute
5R	Recycled Use, Reuse, Reduced Use, Rational Use, and Regenerative Use
10 YFPs	10-Year Framework of Programs

CHAPTER 1

Sustainability Assessment of Policy

1.1 INTRODUCTION

1.2 RATIONALE

1.3 UNDERSTANDING DISCOURSES

1.1 INTRODUCTION

Policy evaluation is a process that measures how far a policy is successful in achieving the goal within stipulated time and cost. Depending on the purposes, policy evaluation may take different forms like: process evaluation, outcome evaluation, impact evaluation, and cost−benefit evaluation (Theodoulou and Kofinis, 2004). Policy analysis on the other hand is done to select the best policy from a set of alternative options. In this respect, policy evaluation is different from policy analysis in the sense that policy analysis is a tool applied before the implementation of policies (ex-ante), whereas evaluation is mostly done after implementation (ex-post) to assess the success of a policy in achieving the target. In addition, policy analysis takes the whole policy unless it is specified, whereas evaluation may take on part of a policy or a set of activities of policies to assess their impacts.

Most forms of policy evaluation are targeted to determine the discrepancy between what was prescribed by the initial policy goals and what has actually been achieved. However, many other forms focus their analysis on different objectives such as: what is the true purpose of the evaluation, how broad or narrow should the scope of the evaluation be, and how should the evaluation be organized and conducted. In this regard, sustainability assessment is treated as an important purpose of policy evaluation. In fact, presently sustainability assessment is becoming imperative for all policies before implementation.

Sustainability Assessment. DOI: http://dx.doi.org/10.1016/B978-0-12-407196-4.00001-5

Sustainability assessment, as we have proposed here, differs from existing approaches of policy evaluation in several important dimensions:

1. Sustainability assessment goes beyond policy analysis or evaluation. It explicitly recognizes that the process by which policies are made has some influence on how the policy is implemented and what the contents of policy are; thereby indicating the likelihood of policy success.
2. It provides a framework and specifies conditions which need to be considered for the integration of society, politics, and economics.
3. It looks at strategic dimensions of sustainable development by integrating policy objectives with the political environment within which they are pursued.

Policy evaluation for sustainability assessment thereby is a tool neither to analyze best option nor to evaluate a set of actions but to evaluate an existing policy in a new way to safeguard the acceptability of the policy while continuing to sustain the resources. Although policies of resource system are usually evaluated to investigate the compatibility or cost–benefit flow to the production of resources, sustainability assessment of a policy is an extended investigation to minimize impacts of resource production/utilization on the environment. Sustainability assessment can potentially help managers to consider a policy situation, its compatibility, and its hidden agenda before implementing it. Thus, assessment for the sustainability of resources and the environment of a nation depends not only on the availability of resources but also on the community influences over the control of resources. In fact, the question of sustainability comes to mitigate whether available resources will meet the demand of the population now and in the future. Policy evaluation for sustainability assessment in that respect, is a systematic approach to researching and exploring the anticipated consequences of policy action and its implementation in regulating people's attitude to resources and the environment. Sustainability assessment thus considers not only the resources or component of policies but also the response or human attitude to it.

The significances of community influences are such that there is a time lag in the occurrence of community attitude/actions and the appearance of their influences on resources and the environment.

The community actions of the present time are likely to reveal their influences to future generations. Moreover, often the influences on resources and the environment due to such actions are irreversible. As a result, the options for future generation to correct changes due to past generation remain limited. Therefore, among other things, the approaches of sustainability assessment provide an opportunity at policy level to negotiate actions expected from present communities on resource and environment that can help to save lives, reduce poverty, and improve the quality of life for future generation.

The community actions and attitudes over the resources and environment of a nation are regulated by policies and legislation as well as by the nature of local and global interactions. In a policy environment, there are beneficiaries, players, and interest groups that play different roles in achieving the goals of a policy. Bringing the examples of past policies, Kumari (1996) demonstrated that narrow concerns of colonials for wealth and power configured the policies of resource control in the past in many developing countries. A sustained supply from the forests of India was of British colonial interest to meet the growing demand for naval and military expansion (Guha, 1989; Saldanha, 1998). Kathirithamby (1998) referred to a similar narrow Dutch and Portuguese commercial interest in the forests of South-east Asia. Those cannot be sustainable options because there was a shift of community intention and interest in resource control of those colonized countries. Indeed, there had been a change in community attitude leading to a catastrophic effect on the sustainability of resources and the environment. Ali (2002) has outlined some of these changes in community attitude due to British policy on Bangladesh forestry. Therefore, it is important that, alongside the cost−benefit analysis, community aspects of policies are analyzed adequately before implementation aiming to reduce the impact on future sustainability of resources and environment.

This book is aimed at outlining some of the elements and considerations of community aspects of policy evaluation in an effort to reduce the future consequences on resources and environmental sustainability. The basic assumption behind it is that sustainability, though oriented to resource and meeting demands, starts from the formulation of policy. Policies are so interrelated that all policies have some roles to play toward sustainability. Therefore, sustainability assessment of policies has an important role in driving total sustainable development.

Not all the elements or considerations for sustainability assessment discussed in this publication are equally applicable for all the societies; however, it is expected that a negotiated selection of factors would be needed to consider what elements should have to be emphasized for a particular approach of sustainability assessment. Before presenting a discussion on elements and considerations of sustainability aspects of policy evaluation, we prefer to present a few paragraphs describing the rationale of study and understanding of the present discourse of sustainability.

1.2 RATIONALE

Sustainability assessment, though a very well known and common term, is rarely conducted for policy reasons. If it is, it is done often for political reasons or to meet the formal conditions of accountability. Typically, the main criteria against which policies are judged are the stated goals of policies that are often vague and reflect sectoral interests. Major improvements in policy evaluation can be made by imposing sustainability as social objectives instead of being subjective to only specific goals of existing policies so that policy evaluation becomes a tool for learning the engagement of policies with other objectives of social development. The current practice is that policy evaluation is often conducted to compensate economic analysis, environmental impact assessment, and poverty and development assessment. Some shortcomings of such practice are that the policy alternatives analyzed in different studies often differ from each other, making systematic comparison difficult and causing opportunities of finding synergies and creative solutions to advanced sustainability objectives to be missed. Sustainability assessment helps to integrate other forms of policy evaluation, thus has potential to contribute in at least in three ways:

1. Integration of different objectives of sustainable development: economic development, poverty reduction, and environmental protection in policy practice.
2. Placement of sustainable development squarely in the policy cycle made up of series of policy actions, from getting a policy problem on the agenda to evaluating the outcomes of the policy.

3. Alignment of policy actions with key components of policy environment—political legitimacy, analytical competence, and institutional
capacity—into the policy process. Sustainability evaluation of policy would help to look into all the issues in an integrated way.

In the present global situation, sustainability can only be ensured
on the basis of mutual understanding of different nations and participation of all components of a society. In practice, this is not happen-
ing. One nation is taking an interest in the resource abundance of
another, cooperating, and collaborating in formulating and implementing resource policies specifically to ascertain their own interest;
whereas in the past, in the colonial instances, nations used to conflict
for resource control. Thus, the outlook for the present concept of sustainability provides a scope for global negotiation and cooperation in
place of conflict. This cannot happen overnight. Mutual respect, political and social changes, and changes in power structure have much to
contribute to achieving an understanding of sustainability. Thus, a policy for national interest only, without considering global cooperation,
is unlikely to sustain—cannot achieve the sustainability goal. Thereby,
sustainability assessment is expected to bring global issues to the consideration of local policies.

The climate policies of some developed countries not ratifying the
Kyoto protocol may be cited here as examples of self-interest.
Sustainability in many developing countries and the present status of
resource production do not reflect the present understanding of environmental sustainability (Williams, 1994). The understanding on environmental requirements for production systems is more intense in
developed countries than developing countries. However, poverty and
ignorance play a salient role in the resource production system of
developing countries. Sustainability assessment of policies is expected
to reveal such differences in understanding and can lay a roadmap to
mitigate the sustainability problems of different countries. This means,
if sustainability assessment becomes obligatory, global policies could
have the power to persuade local policy makers to include global sustainability concerns in their considerations.

Through sustainability assessment, illustrations on the social and
global cognitive bases of the way the sustainability problems are constructed in a community are expected to create some interpretation of
social processes that may reduce sustainability problems and may

produce factors through which the actors can be mobilized around certain concepts or ideas of common understanding of sustainability problems. Sustainability problems may originate from mutuality of intersocial resource transactions (referred as venture problems) and/or mobility of intrasocial production (referred as maverick problem). Mobilizing actors around "venture" or "maverick" problem requires enlightenment on the cognate bases of the society in question as well as regional or global integrity.

Venture problems that justify presenting community responses from other societies/countries require a cognition on general the nature of the problems. However, in some societies, actors may be found to move from "venture" to "maverick" status depending on the issues of environment and the cost–benefit involved in it. For example, actors from the USA are very concerned and usually consider the issues of environmental sustainability as venture problems, but with the question of emission cuts (e.g., in Rio and the Kyoto protocol), their attitude appeared more "maverick," influenced by the interests of their own state politics. Thus, inclusion of background issues of communities of a country and/or negotiation of communities of different countries in sustainability evaluation of policies cannot be seen as an entity free from the influence of actors originating from within or across the society. Thereby, explanations of social factors are considered as import components of the sustainability assessment of policy.

In resource and environmental problems, influencing factors may originate from spatial scale (many countries) or temporal scale (past to present); however, contextualizing those factors with society and resource remains an essential component of policy evaluation. Policy evaluation accompanied with the contextualization of problems in spatial and temporal scales may be more convincing to transform a specific issue into a general issue of sustainability. In support, to discuss the contextualizing of issues in a wider scale, a brief discussion is made here for an understanding of discourses explaining why and when sustainability can be a "venture" or "maverick" issue (involving spatial and temporal scales).

1.3 UNDERSTANDING DISCOURSES

From the discussion in the previous paragraph, it is apparent that an understanding of resource and environmental principles is important to identify/transform global initiative on particular issues of a local resource policy. On the basis of approaches and requirements of sustainability assessment, the relevance of a policy to sustainability discourses of resource and environment may be ascribed on five assumptions:

1. Resource and environmental change is not a temporary phenomenon but is structural in character. Those changes are accompanied by series of problems.
2. Resource and environmental problems have their own meaning associated with the prevailing social order. Douglass (1988) defined destruction of resources and environment as removing them "out of place," which does not appear to be always true. Things may be portrayed as "out of place" for limits of adoption or adaptation as well. When changes in resource and environment within societies become unacceptable, the changes could be treated as problems.
3. Debates on nature of resource and environmental degradations reflect the contradictions of social developments. Reducing degradation is often considered against the idea of social development, on the other hand, the present mode of social development creates degradation. These sawtooth debates create an impression that resource and environmental degradations are parts of the social development process.
4. The problems of resource and environment can hardly be discussed in their full complexity at a particular time. For example, in the USA the issues of deforestation arose in the early 1900s, soil erosion in the 1930s, pesticide pollution in the 1960s, and resource depletion in the 1970s (Hajer, 1995). Although the problems were related to forest land use, all the problems did not occur at the same time.
5. Resource and environmental issues are discursive and span several branches of knowledge; as a result, addressing all the issues in a single policy is very difficult.

These assumptions of resource and environmental discourses show that understandings on social, political, and physical backgrounds of different countries are important for clarifying the sustainability problems as "venture problem." The discursive nature of the subject "resource and environment" emphasizes that a thematic background should highlight the particular aspect of policy; otherwise, policy evaluation will be a confusion of resource and environmental discourses. While the physical attributes of resource and environment could be the same for a region, social aspect of problems may vary from nation to nation. Therefore, the course of sustainability assessment of resource and environmental policy should follow a distinctive and relatively precise path avoiding physical contradictions.

At present, involvement of politics with resource and environment also gives a variation to resource and environmental policy from country to country. One of the difficulties of sustainable policy design is that it is not a problem closure, that is, there is no common set of socially acceptable solutions for all well-defined resources and the environmental problems; rather it is an interpretative activity. Problems may remain somewhere else. The political emblem over resources and the environment often makes the interpretation contradictory to policy activity for some political gain of different societies. On the other hand, formal policy is a state relevant issue; thus, it becomes circumscribed by politics often involving actors from a global level.

Although the actors from global level are neutral, they often induce policy. They may or may not act as consultant or advisory service to the national governments (e.g., experts, world media like BBC), but their reporting induce political actors directly or indirectly to activate policy actions. These actors have been termed in this book as "contextual actors." There is another group of global actors who donate and/or financially get involved in the national resource and environmental issues; they thus indirectly dictate political actions and are policy actors (e.g., donor agencies and multinational corporations). These actors are indicated in this book as "extraregional" actors. However, as politicians try to maintain an image of being in control, contextual or extraregional actors may proceed through critiques and administrations that may conceal information for avoiding the status of policy failure. Thus, social and political actors and their coalition with other (e.g., contextual and extraregional) actors may transform a venture problem to a maverick problem and vice versa.

Outcome of a policy may be formed from the "discourse coalition" (based on merits of the policy) or conventional "political coalition" (based on followers of past politics, e.g., colonial policy). The impact of both forms of the policy coalition cannot be the same. Sabatier (1987) argues that policy change can best be explained as a demand of a competing "advocacy coalition" at the level of policy system (organizations that are concerned about policy) or a demonstration of corresponding "provoking coalition" at the level of a social system. Present advocacy on the demand of sustainability of resource and environment, as a worldwide paradigm, is expected to have impact on policy change if adjusted with other coalitions. Thus, the sustainability in a policy process can be seen as a struggle of coalitions, which largely depends on the respective social traditions of actor coalitions. Coalitions occurring in such a way result in social learning, which may be called "cognitive learning," concerning the perceived influence of external dynamics (interest groups) and construction of the past (time influence).

There could also be other sources in the real world from both outside (e.g., changing socioeconomic parameters and changing coalitions in government) and inside the subsystem (e.g., changing the personnel), which help to bring about policy change (Palumbo and Hallett, 1993). Breakdown of communism/socialism can be a good example here. Berstein (1991) suggests that the policy of political forces within a given domain of social policy is the most important determinant of the way in which substantial policy issues are dealt with. Sustainability issues within the policy are also required to have different dimensions of political coalition to function.

Hajer (1995) explained that structuring ideas generated in cognitive learning and putting those to work in the close interaction that takes place between various actors of a coalition within the social domain remains hidden. Therefore, while dealing with the analysis of policies, evaluation of political interplay in such a domain may not be possible through the analysis of a distinct coalition. However, controversies between those coalitions should always be understood against the background of external parameters, like institutional structures, social structures or geographical predisposition, or even indeed the economic dependence of governments.

Sabatier (1987) presents the advocacy coalition as an analytical alternative to the analysis of institutions and actors, but when too many organizations are active in a subsystem, identification of the role of an institution is not very simple. However, in the advocacy coalition, there may be some behavioral elements, e.g., what people say may differ from what they practice. Similar differences are likely to be available among the individuals engaged in policy making. Looking at the belief of individuals as variable, explaining policy change would be very difficult for a sustainability interpretation; but taking over the "rival coalition" (contextual, extraregional, or antigovernment coalition) may provide an alternative perspective of analysis. Considering the condition of different issues of policy and sustainability, policy evaluation can be justified when the rival coalition is taken from the arena of development and growth.

This book has been organized anticipating that policy evaluation is the key factor for communicating the policy discourses and policy aspirations to meeting the sustainability expectation of future generations. However, limitation of the size of composition also needs to be kept in mind. Hence, this study excludes detailed presentation of examples on issues relevant to resource policy and environmental sustainability; rather introduces the foundation and general principles of the specific discussion.

The aspects of variation of sustainability issues of resource and environment have been covered on the basis of socioeconomy, culture, and the differences pertaining to geographical influence. The ecological characteristics of a country depend on the geographical position. On the other hand, the dynamics of ecosystem are usually influenced by the demographic and resource use characteristics. Besides, presently culture and resource use are influenced by the globalization factors, whereas, in the past they were affected by colonization. Both colonization and globalization and their relationship are relevant to build the foundation for sustainability evaluation of resource and environmental policy. This chapter highlights the relevance of those points along with these sustainability issues.

Chapter 2 briefly introduces the surroundings of a policy (climate), national, and international, which influence the policy formation and formulation. The sociopolitical changes within the society and the influence of cross-national commitments may have a bearing on the resource policy statements but often the implications, especially in the developing countries are different, the consequences of which may bring an adverse result to the sustainability. A few of these cases have been placed in Chapter 3 to produce a background for the reasoning of sustainability analysis of policy and for recommending the future likelihood of divergence and convergence of the goal of resource policy.

The discussions have been forged under the factors of policy input, policy process, demographic factors, social and political institutions, and present and future resource standings of a country in question. The general and unique causes have been filtered out, with an expectation that formulating or improving resource use policies on the basis of native problem and/or common problem would be much easier. A few of those points have been highlighted in subsequent Chapters—4–10.

The background of deviation from the goal of environmental sustainability can be related to the demographic and resource use issues of a country. At the same time, there could have been involvement of many organizations and/or institutions, interested in resource and environment for accentuating the causes of divergence and convergence of problems. Hence, the issues of all chapters have been presented under different subheadings, such as demographic issues, resource issues, basic land use principles, and paramount environmental movements.

After a comprehensive review of available literature on policy evaluation, a brief description of the tools for resource and environmental evaluation has been presented in Chapter 11. The rationale for the choice of method/tools for analyzing the resource use of policies in question is also explained in this chapter. However, investigation of background of policy evaluation is expected to highlight difficulties. The common tradition of policy evaluation is to verify their achievements on the basis of economic principles, which cannot meet the requirements for the evaluation of environmental sustainability. Chapter 12 presents the limitations of sustainability assessment of policies in terms of understanding of policy elements and boundaries. Chapter 13 includes a conclusion of the study showing the key features and key aspects of sustainability assessment of policy. A few suggestions have also been given ascribing the future resource use policy for environmental sustainability.

CHAPTER 2

Sustainability Climate of Policy

2.1 INTRODUCTION

Although there are volumes of discussion on sustainable development and the environment, sustainable policy merits a special attention. Policy sustainability can be considered as a kind of new connection to all efforts of sustainable development. A sustainable policy designates the integration of goals and activities of a policy with sustainable development. Therefore, sustainability assessment of a policy can be considered as a process that helps policy managers to integrate the objectives of sustainability into policy actions in a given sociopolitical environment and to plan a strategy for policy implementation. The sustainability of resource and environmental policies depends on socioeconomic and sociopolitical issues considered in planning and management of a resource system to meet the needs of society. Perhaps, that is why different policies attempt different strategies for the improvement of resource use and to ensure sustainability. However, social attitude like corruption, nepotism, and nonprofessionalism of policy actors

Sustainability Assessment. DOI: http://dx.doi.org/10.1016/B978-0-12-407196-4.00002-7

may cause policy failure in many instances. Thus, performances of resource policy largely depend on social factors, sociopolitical actors, and resource sectors of a country. These factors and their interplay within the society constitute the sustainability climate, an understanding of which is important for policy evaluation. This chapter essentially aims at presenting sociopolitical elements that create a congenial climate for sustainable policy.

The factors of sustainability climate are not exogenous; rather, they evolve along with the society. A good understanding and explanation of policy issues addresses the social activities as well as the processes of modernization. As a result, rationalization of resource use and negotiation of social contradictions are important issues for sustainable policy. The fundamental way of looking at those processes is to select different forms of ecomodernist tools and practices, which can be used in mitigating social contradictions but which have not been given importance in the policy action. Thus, understanding the economic processes, social dynamics, development progress, and institutional characteristics is important for general clarification of policy climate.

The problems of definition as well as solution to sustainability should be socially acceptable and technically viable. If an environmental problem is addressed by a technically efficient but socially insensitive policy, that policy may result in regulatory failure. There are certain policy instruments that need to be utilized appropriately for solving the problem and for the viability of policy. For example, an intervention or a nonintervention method, if adopted to prevent social construction of an environmental problem, may not help in achieving environmental planning for sustainability. The social construction of a problem should be prevented by social measures not by intervention measures. According to Hajer (1995), the social climate of a policy problem largely depends on the following three dimensions:

1. *Discursive closure*—delineates which aspects of the problem are included and which aspects have been left out from the policy.
2. *Problem closure*—which shows whether the strategy of regulation helped to achieve certain goals of sustainability.
3. *Social accommodation*—gives the understanding of any particular way the problem has been positioned or priority it has been given.

These three aspects of a policy may or may not be supportive to each other. However, in environmental planning and sustainability problems, these elements may show a relationship to each other and can be correlated to a particular resource status targeting to sustainability. If a policy is not "problem closure" for a resource, then it can be seen whether it was "discursive closure" or was "socially accommodated." Thus, the basis of sustainability assessment of a resource policy will be sound if the resource dimensions can be coordinated with social dimensions.

In many cases, policy formation of different countries shows the dominant form of regulation for controlling the resource use. Most of them have given rise to the environmental changes because they were not socially accommodated. Thus, an understanding of policy problem is related to the society addressed by the policy and the way problems of resources are addressed. Therefore, before focusing on sustainability themes, it is important to develop an idea about the components and ingredients of a policy and their relationship with the society. This chapter is aimed at presenting a review on those aspects of policy. At first, the views are organized and presented considering the emergence of a sustainability concept, which is the main target of policy evaluation. Then, the basic concepts and basic considerations of policy evaluation are reviewed. Policy-related social elements are also considered before presenting specific elements of policy climates important for assessment. The specific points of resource issue and their possible relationship with the environment, market, and other policies of the society are also explored.

2.2 EMERGENCE OF POLICY SUSTAINABILITY

The evolution of ideas of sustainability in a society may have strong linkage with the evolution of policy climate and emphasis of policy target. Norton and Noonan (2007) have given an account on the evolution of the discipline with considerable debate; not only on what constitutes sustainable development but also on what the term itself means. Although the World Commission for Environment and Development (WCED) has brought the concept of sustainable development into the limelight in 1987, the sustainability issues were first undertaken in global negotiation in the 1992 Earth Summit of Rio de Janerio outlined under agenda 21. Following the summit, concerns around the impacts of unsustainable consumption and production activities have led to the call for the establishment of 10-year framework of programs (10YFPs) in the Johannesburg plan of implementation, agreed at the 2nd World Summit 2002 on sustainable development. The aim of the 10YFPs was to accelerate a shift toward Sustainable Consumption and Production (SCP), thus promoting social and economic development within the carrying capacity of ecosystems by integrating economic growth with environmental improvement.

The Marrakech Process on SCP was developed under the 10YFP on SCP to be a global framework for action on SCP that countries can endorse and commit to in order to accelerate the shift toward sustainable consumption and production patterns. Although the global emphasis on sustainability management is quite strong, the policies of different societies are not equally committed to maintaining the resources and environment. Policy climate is thus expected to reveal an understanding on ambiguities of resource and environmental sustainability. The following section presents the factors and issues associated with the emergence of sustainability ideas in general. Subsequently, those factors and issues will be considered within the elements of policy climate for outlining the principles of sustainability assessment.

2.2.1 Population and Resource

The understanding about the impact of population growth on environmental resources and human welfare issues are as old as beginning of civilization. Dietz and Rosa (1994) cited references of a famine of the fifth century BC and how population outfaced the production of food. The idea of linkage between population and resources was conceived in more concrete form in the eleventh century, when population enumeration used to be done to allocate the resources on priority areas. Nevertheless, it was not until 1798, when Robert Thomas Malthus (1766–1834) published his famous work *Essay on Population*, that the systematic relationship of population and resources was addressed.

The debate continued almost for a whole century as "present and predicted competition of species for resources" until nineteenth century when Charles Darwin (1859) developed his theory of evolution *The Origin of Species*. He described the competition within and among species for resources and environments under the theory "survival of the fittest," showing how the species survive and evolve. The discoveries of Darwin first brought the science into the sustainability concept. Subsequently, social sciences and biological sciences were looking into sustainability as a common problem and many explanations of Malthus' proposition have been tried in the light of Darwin's theory.

This debate has been addressed in the social sciences as a "human welfare issue" of policy for two centuries with the help of "welfare economics," mainly in the form of distribution of resources. Involvement of biological sciences flourished in the discourse of ecology in this line. Thus, social modernization and scientific progress blended together producing a new reflex of policy target. The issues of sustainability thus developed from social problem to policy problem. In practice, policy confusion of environmental sustainability remains how to address the boundaries of economic and environmental welfare of human society. However, the word "sustainability" was brought to the policy attention only recently (Holdren et al., 1995), within a couple of decades, and spread not only to environmental scientists but also to other actors like activists, economists, social scientists, and policy makers. Thus, the policy outcome of sustainability became dependent on how the social institutions and organizations were modernized to coordinate and accommodate the roles of those actors in the society.

2.2.2 Modernity and Sustainability

Modernity started from the age of the industrial revolution. While industrial revolution is about 200 years old, the sustainability, environmental sustainability in particular, has been seen as a problem only in about the last three decades. On the other hand, while the industrial revolution happened in few societies, sustainability is an issue of all the societies. Therefore, modernity and sustainability did not occupy the same position in the policy. At the same time, the prophecy of global doom (Meadows et al., 1972) enlightened different aspects of reality and was discussed in different issues of sustainability of resources. The fear of nuclear power started a new era of environmental movement in 1970s and mass demonstration started proclaiming for a safe environment. Thus, within a short time the sustainability paradigm received a global dimension on the wing of environmental stability. Therefore, the dynamics of policy climate remains limited for modernity but global movement is accelerated for sustainability.

The environmental movement in the 1980s stepped beyond the mass demonstration and became more radical and practically policy oriented (Hajer, 1995), which in turn has brought the new incentive in the resource policy process with a change in the political strategies and organizational structure. However, the constitution of environmental movement in 1980s started changing due to direct involvement of the secondary international institutions. Almost in the same period of political cold war, sustainability movement of the states became techno-biased. However, it is undeniable that the origin of street demonstrations on environment in 1970s has awakened politicians and policy makers to develop sustainable policy discourses in the society.

There is a spatial variation among the societies as to how the transition to sustainability has been pursued (Grainger, 1999). Different societies pursued the transition in different way. While in developing countries, regulatory instruments got preference for ensuring a policy environmentally worthy, in developed countries, organizations like OECD were in favor of economic and fiscal instruments (OECD, 1984). Eventually, two schools of thought have developed around the issues of sustainability, both having merits and demerits of their own.

In certain aspects, however, the anticipatory policy measures (this and that will happen) did not go further for environmental sustainability because technological innovation has always meant that the anticipation was incorrect and political motivation did not like the measures to be taken. Instead, several publications like *Facing the Future* (Interfuture, 1979), *Our Common Future* (WCED, 1987), *North−South: A Programme for Survival* (ICIDI, 1980), and *North−South Common Crisis* (ICIDI, 1983) have changed the sustainability concept toward mutual cooperation in the field of environment. Thus, the transfixing of words "environment" and "sustainability" as a term "environmental sustainability" could have been caused by few factors:

1. Substantial advancement of the scientific understanding about the consequences of environmental change.
2. The leading countries of the world became free from cold war phenomena and took the leadership in environmental sustainability.
3. Paying world's attention to the behavior of the developing world for meeting their basic needs.
4. An avalanche of reports and subsequent works on sustainable development.

Thus, the popularity of the term "sustainability" in the policy climate was caused by academic productions of contesting schools of thought and the unified leadership given by the developed world.

2.3 CONCEPT OF SUSTAINABILITY

A comprehensive idea of sustainability has been introduced by the WCED in 1987 as the considerations of actions in managing resources and environment so that they meet the needs of present generation without compromising the ability of future generations to meet theirs. It embraces two key notions:

1. The concept of needs to which overriding priority should be given.
2. The idea of limits to the environment's ability to meet present and future needs.

Since the inception of Brandtland concept of sustainability, at least 40 working definitions of sustainable development have been identified by Hajer (1995), but no definite framework was defined to say something is sustainable and something is not. As a result, sustainability targets the policies facilitated by debate and ambiguity. One part of the debates is on discourses of societies which can successfully manage their environmental resources over a long period of time. The pursuit of sustainability remains in balancing acts involving implementation of policies, strategies, programs, and projects that treat resource and environment as a single issue. It also demands changes in the mindsets, attitudes, and behaviors of stakeholders. Thereby, sustainability has been expressed on the basis of relative exhaustion of the factors that determine the abundance of resources and their footprint on nature. However, certain useful concepts have been outlined in Geyer-Allély and Eppel (1997) for expressing the performances of environmental sustainability. They are presented in the following sections.

2.3.1 Steady-State Economy

The concept of steady-state economy assumes that economy is an open subsystem of earth's ecosystem, which is finite, nongrowing, and materially closed. The steady-state economy is thus a nongrowth economy in biophysical equilibrium where:

1. Stocks are maintained at a sufficient level for abundant life of the present generation and are ecologically sustainable for a long time.

2. Service is maximized given the constant stock.
3. Throughput is minimized.

This is a version of "real economy" in which the economy is guided by the stock of diverse resources and symbolizes a sustainable situation. Real economy rejects the commoditization of nature other than to meet the subsistence needs. Thereby, steady-state economy demands that the existing corporate economy needs to be dismantled and a new one, founded on environmental rationality, needs to be erected in its place to achieve sustainability.

The assertion of steady-state economy stems from the realization that the root cause of sustainability crisis in resource and environment is to be found in an economic process that operates as an entropic force and is driving the planet rapidly to its death. Moreover, with the current economic structure in place, it is not possible to slow down the growth as this structure stimulates economic expansion, increases the consumption of nature, and causes destruction of bases of sustainability of both the economy and life. If these assertions are correct, then they put into question the possibility of reconstructing the economy by incorporating environmental policy, technological innovations, and distributive balances. In other words, steady-state economy assumes that it is possible to rebalance the economy within the theoretical and instrumental rationality of policy.

A steady-state economy is opposite to boom and burst economy and designates a sustainable situation of economy. If resource flow is not continuous, it is unlikely that an economy will remain in steady state. If policies are perfect, it is likely that resource flow will remain continuous from a renewable system. From a nonrenewable resource system, good policies maintain flows of resources for a long time, for example, through 5Rs—reuse, recycled use, reduced use, regenerative use, and rational use. Thus, the implication of a steady-state economy is to indicate the policy circumstances of sustainability on resources and environment.

2.3.2 Carrying Capacity

Carrying capacity is a quantitative concept that assumes the limit, though difficult to estimate, of the ability of natural ecosystem to support continued growth of population within the limit of abundance of resource and within the tolerance of environmental degradation. The size of population that the carrying capacity of a resource system can support mainly depends on the size of the needs of that population. The size of the need cannot exceed the limit of carrying capacity to maintain sustainability. Although the definition of carrying capacity has been forwarded from natural science, its orientation to the needs of population converts it to very much related to management and policy issues. Key factors for manipulating needs are population number and density, affluence and technology, depletion rate of renewable and nonrenewable resources, and the build up of hazardous wastes in the environment. Therefore, understanding the carrying capacity concept is important for formulating sustainable policy.

While carrying capacity is relevant to natural condition of resource, environment, and communities present in the system, human beings' can influence the carrying capacity of a system significantly through harvesting resources of their interest and through maintaining the relationship among the communities present in the system. How the system will continue largely depends on how the human being treats the system for the derivatives of his/her own well-being. The comparative assessment of carrying capacity over a spatial and temporal scale determines human beings approaches to the system. Because policies regulate the attitude of human beings to a particular system, thereby comparative assessment of carrying capacity is likely to provide information on the nature of influences of policies on resource sustainability.

2.3.3 Ecospace

Ecospace is a concept based on quantitative limits of carrying capacity and critical load set on the basis of scientific analysis and political evaluation of the risks associated with policy not to exceeding the sustainability limits. This is also associated with resource distribution limit that suggests an allocated ecospace at a national, regional, or per capita level (on the basis of global fair shares). This concept is important for understanding resource sustainability.

Using the ecospace concept, Wackernagel and Rees (1996) have developed an index calculated on the basis of distribution of resources in different nations. They have considered "ecological footprint" and available "ecological capacity" to calculate ecological deficit (the difference between the two). The findings can be considered as the ecospace situation of different countries. Ecospace is different from carrying capacity in a sense that, in addition to quantity of resources, it signifies the distribution of resources. For example, if a species can thrive across a country or a region, availability of such species in a particular location would be a risk to sustainability. The ecospace of that species is likely to become limiting for its continuity. Because policies may likely to have impacts on the distribution or recovery of resources, therefore, ecospace situation of a particular resource is likely to indicate the policy consequences of that society.

2.3.4 Ecological Footprints

According to Wackernagel and Rees (1996), ecological footprint measures the consumption that requires estimation of natural capital requirements based on the interpretation of carrying capacity. It takes into account the impacts of technological development and trade. As wealth and consumption levels increase, so do the area of productive land, ecological footprints (EF) and throughput of material (ecological rucksacks, ER) required for supporting every individual. A key assumption of EF is that technology and trades do not expand the earth carrying capacity in the long run but only displace geographically the effects of increased consumption levels.

The concept of EF is not on the basis of mere ecological productivity but an attitudinal sanctity that preserves nature and that links resources as the source of life and production to ethics and esthetics of nature. It entails moving from a mystification and adoration of nature to an awareness of our human culture and an ethics of responsibility toward living creatures. The sustainability in EF can thus be treated as a reconstruction of economy into a process of resignificance of life and human existence. Building sustainability through manipulation of EF is not, then, simply a way of ecologically managing nature's productive potential. It entails a reappropriation of nature in its own relational splendor with other components of nature, not just at its productive valor. Thus, in association with technological ability, the importance of cultural creativity of human beings cannot be seduced to achieving efficiency and sustainability. Cultural creativity has to do with the meaning and the values ascribed to nature as a territory of life and as a space for recreating culture.

EF also indicate the influence of policies beyond the resource area. Policies of one resource area may have influence on other resources. Even policies of one geographical area may have influences on other geographical areas. Dams on international water channels, international trade policies can be mentioned as few examples. Policy evaluation may deny such influences, but sustainability assessment may not; because sustainability assessment has spatial and temporal jurisdictions. EF identifies the sustainability situation of policies within and across the resources at local or regional or cultural measures.

2.3.5 Natural Resource Accounting/Green Gross Domestic Product

As the traditional measures of aggregate income do not accurately reflect the welfare in economy (Adger and Whitby, 1993), it is not possible to reflect environmental sustainability by economic accounting only. The concept of natural resource accounting can be seen as a tool for demonstrating linkages between the environment and the economy to correct distortions in standard measures of national "growth" and "welfare." Natural resource accounting dictates strong or weak sustainability that depends on critical material capital. Green Gross Domestic Product (GDP) measures are based on quantitative indicators of national performance based on data relating to the availability and use of natural and environmental resources (stocks and flows) and incorporating qualitative judgments as to what constitutes economic, environmental, and social welfare.

The concept of natural resource accounting and green GDP is a rationality that incorporates thoughts and values, reasons and meanings, that is open to differences and diversity, that seeks to break down the unitary and hegemonic logic of the market to build an economy based on specific relationships between resource and environment and is thereby symbolic between culture and nature. The development of green GDP is at its infant stage and is not usually practiced. However, comparing green GDP with total GDP of a country would reveal the effect of policies on natural resources. It might be that the segregating effects of specific policies may not be likely from such an approach; however, correction or accommodation of specific policies is possible to improve the situation in green GDP.

2.3.6 Ecoefficiency

Ecoefficiency is a management strategy based on quantitative input–output measures which seeks to maximize the productivity of material and energy inputs in order to reduce resource consumption and pollution/waste per unit output and to generate cost savings and competitive advantage. Ecoefficiency of policies indicates how a policy influences in reducing waste, recycling materials, reducing pressure on inputs, optimizing output, and other principles related to 5R principle.

The ecoefficiency concept tries to mitigate sustainability between economy and ecology and gives rise to concern over the feasibility of alternative production rationality founded on different production principles and social values. Concepts based on ecotechnological productivity, production with ecological rationality, and cultural creativity are the results of ecoefficiency considerations in the production system. This conjunction of ecoefficiency promotes ecological conditions that underpin sustainability and tries to substitute the economy founded on capital, labor, and technology as the basic factors of production.

Thus, policies that discourage externalization and devaluation of nature and encourage natural resources to propagate while feeding the process of production promote ecoefficiency. Such policy outcomes to the environment originate from social contributions like training human beings, investing in technology, motivating community people, and even valuation of resources. Thereby, ecoefficiency is an important social determinant of resource policies that indicates sustainability.

2.4 SUSTAINABILITY INITIATIVE

Global initiatives in resource and environmental sustainability are trying to formulate and integrate conservation and utilization principles of different resources through social policies. The emphasis of global initiatives depends on the nature of problems. This study conceptualizes the problems in following two ways:

1. Some problems are global (e.g., greenhouse effect), and if they happen, will affect all communities. Moreover, a solution to those problems is not possible by a single country or a few mighty countries; all the countries need to give concerted effort to solve those problems. In this book, those problems are termed as "venture problem" (note that nuclear catastrophe can also be a global problem

but the solution remains at the hand of few countries, thus is not a
venture problem).
2. Some other problems are local in nature (e.g., fuel wood crisis) and
   the solution remains in the hands of local authorities. Such problems have been indicated in this book as "maverick problem."

Understanding the maverick and venture problems are important
because they can make the difference in cognition of the problems
in national policies and the global initiatives. The national policies
may tend to solve the "maverick problem," whereas the global
initiatives may target the "venture problem." Environmental sustainability will be hampered if such contradiction cannot be avoided and
if "venture" problems are not solved through addressing "maverick"
problems.

Another problem faced by the developing countries is that under
the consistent pressure of global needs, the trend of resource and environmental policies may get shifted so fast that many developing countries will not be able to keep up with it. For example, forest resource
organizations of newly independent countries in the 1950s and 1960s
were not even able to shed the colonial coat they had inherited from
colonization. Their institutional build-up was also immature in
adopting changes. Thus, when transformation of a global system is
more inclined to corporate economy and privatization of entrepre-
neurs, developing countries are likely to caught up in "maverick
problems;" consequently, they may find themselves unready to
adopt the global changes. This problem of nonpreparedness also
causes substantial impediment to changing the legacies of the past
and implementing the global initiatives—often compounded by
social tradition.

The initiatives of global changes are driving nations and countries
with an accelerated change in the resource and environmental policies,
to have apparent sustainability. However, Wilson and Bryant (1997)
and Daniere and Takahashi (1997) stated that the developing countries
are adopting a rapid change in the land use contributing at least two
conflicting goals:

1. Increasing effort to modernize and promote growth.
2. Improving environmental standard.

Waggener and Lane (1997) express concern that industrialization and urbanization are playing major roles as drivers of resource use change but not in raising environmental standards. Multinational companies are involved with cash crops like rubber, cocoa, coffee, and palm oil. Thus, the expansion and intensification of industrialization, urbanization, and agriculture in developing countries are pushing the environmental claims to the status of peripheral and optional. Concomitantly, there has been a rapid shift from a predominantly rural-based population to one that is now increasingly concentrated in large cities (Lebel and Steffen, 1997). The conservative traditional lifestyle is rapidly changing to a consumer lifestyle. As a result, pressure is also increasing on resource sustainability.

To meet both the ends of growth and environment, devising a far reaching resource management policy is a national priority of many countries (Dias and Begg, 1994). But the nature and implication of environmental sustainability and resource use policy are not simple. They involve social, political, economical, ecological, and even spiritual insight. Thus, Valadez and Bamberger (1997) defined evaluation as:

> the internal as well as external management activity to assess the appropriateness of a policy design and implementation method in achieving both specified objectives and general objectives and to assess the results, both intended and unintended, and to assess the factors that affect the level and distribution of benefit produced.

Although the subject matters of sustainability individually merit thorough discussion, the bridge between resource and environmental sustainability need to be investigated within the policy as widely as possible. However, within the limit of the purpose of this text we recommend following a "9C" arena for evaluating diverse components of resource and environmental policy:

1. Cause assessment (why the policy was required).
2. Condition assessment (how the policy was formed and implemented).
3. Component assessment (what are the components/institutions involved?).
4. Content assessment (what are the contents of the policy, what would be effective for achieving its goal).
5. Context assessment (where/under what circumstances the policy could be implemented).
6. Conflict assessment (what or who are likely to contradict).

7. Contrast assessment (difference of policy in writing and policy in practice).
8. Cultural assessment (how the policy regulations are likely to be observed by the social element).
9. Consequence assessment (what would be the likely outcomes/ impacts).

These outlines may not be exhaustive but illustrative. The scope and coverage of them in relation to evaluation of resource and environmental policy and options of sustainability could be partially overlapping. However, detailed studies on those issues would be useful for resource managers to implementing policies in a sustainable manner. A list of actions that may reveal the above 9C qualifications of sustainability assessment of resource and environmental policy is outlined here:

1. Comparing how the policy processes influence the resource and environmental decisions.
2. Exploring the relationship between people and environment defined within the respective resource policy.
3. Distinguishing the mode of people's well-being and measures of environmental sustainability addressed in the policy.
4. Studying how the equity issues have been addressed.
5. Illustrating how the linkages of poverty, income generation, and resource activities have been established by the policy.
6. Finding out the implications of community factors (e.g., ethnicity, gender distribution, and education) in explaining the environmental sustainability addressed by the policy.
7. Contrasting the influence of tradition, culture, and modernization on environmental sustainability of resource use.
8. Knowing how the technological progresses and their implication in resource and environmental sustainability can be considered within the policy.
9. Examining how particular economic and political aspects relevant to environmental sustainability and land use policies have been highlighted.
10. Considering how political ambition and the influence of lobbying groups affect the policy outcome.
11. Evaluating the comparative perspectives of the role of different actors as they relate to the sustainability of resource and environmental sustainability.

12. Analyzing the influence of market interest, market value, and resource trade on the sustainability issues of resource and environment.
13. Monitoring how policy influences the growth and relationship with sustainability.
14. Assessing the extent of national resource and environmental issues covered by the relevant policies.
15. Investigating the impact of group or regional issues on the sustainability of resource and environment.
16. Searching how the global initiatives, conventions, and prescriptions have been incorporated and influence the sustainability of national resource and environment.
17. Highlighting the relevance of development characteristics and policy implementation against the resource and environmental sustainability.
18. Synthesizing how policies influence the availability of different forms of resource and environmental opportunity.
19. Relating the present versus traditional resource use and their prospects to environmental sustainability.
20. Estimating the positive outcomes of resource and environmental conservation issues addressed in the policy and their implication to sustainability.
21. Verifying the relationship among different policy frameworks, instruments, and their implication on resource and environmental sustainability.
22. Explaining the scope of policy contexts with resource and environmental sustainability.
23. Showing how the structure and process of decision-making and policy implementation affect the outcomes of sustainability.
24. Demonstrating whether other external policies have any undue impact on the sustainability issues relevant to resource and environmental policies.
25. Identifying the role of organizations, institutions, nongovernmental organization (NGO), public and private sectors in policy formulation and implementation.
26. Appraising how incentives and appreciation influence policy implementation.
27. Ascertaining how tenure and financial security are useful for policy implementation.

This list of actions is constructed in the light of the important roles of policy on resource use and environmental sustainability. There might be additional steps needed or some steps may be subtracted depending on the type of resource and extent of environmental sustainability targeted. Additional issues may also need to be considered if the resources are technologically unique and skewed in territorial distribution across the world. There could be social implications of resource and environment as well, the nature of which depend ultimately on the broader perception of the possibilities of "modernization" of developing countries.

Based on forest resources Bryant et al. (1993) highlighted further an extreme situation; that the ecological consequences of a forestless country would not be necessarily cataclysmic. The common idea is that economic development is enhanced through the efficient use of natural resources, thereby, natural resources like forests have been the mainstay of many developing economies. Even if the environmental costs of deforestation are factored in, disciples of the modernization school might still argue that forest depletion is the most sensible long run strategy for development. Supporting the idea, Summers (1992) wrote:

> we can help our descendants as much by improving infrastructure as by preserving rain forests, as much by educating children as by leaving oil in the ground, as much by enlarging our scientific knowledge as by reducing CO_2 in the air.

This concept of modernization, though encouraging for efficient use of resources, will hardly be effective for environmental sustainability. Such views do not encourage the value of tradition and culture. Thus, specific principles and considerations of policy evaluation will be pursued through enquiring the rationale of shifting traditional and cultural values influenced by resource and environmental policies for modernization.

CHAPTER 3

Characterizing Sustainability Assessment

3.1 INTRODUCTION

Sustainability assessment characterizes the policy influences on the status of resource and environmental conditions. Sustainability assessment could be addressed partially by looking at the recorded information of the peoples' responses to policy actions. Policies are complex formulations, have relations with other policies, manifest resource consumption, and have long-term perspectives. Over time, there could be many deviations of issues and situations under which the original policies were formulated. Such deviations in policies need correction. The corrections are usually made on the basis of feedback analysis on policies. Therefore, policy evaluation is not a one-time process of predicting probability of achieving goals at relative ease but a continuum of analysis for the correction and appeasement of actions to avoid policy failure and to ensure sustainability.

Sustainability Assessment. DOI: http://dx.doi.org/10.1016/B978-0-12-407196-4.00003-9

Usually policies are assessed for cost–benefit analysis which is a traditional approach of economics. However, social sciences, on one hand and biological and environmental sciences on the other have addressed the issues separately, and often antagonistically. If examined closely, the term "sustainability" till now is limited within the claim and concern of certain issues (e.g., the rain forest movement). A real push into the era of policy sustainability has not yet been reflected in the resource and environmental sectors of developing countries. Many issues of society and sustainability need to be internalized such as tradition and culture, population, resources, and pollution for a comprehensive program of policy sustainability. The characteristics of a sustainable policy can be assessed through understanding how the inclusion/exclusion of these issues in the policies could have a bearing on the sustainability of resource and environment of developing countries. Some of the issues are described in the following sections.

3.2 RESOURCE SYSTEM

The first point of characteristics for assessing environmental sustainability is embedded within the natural resource system and its relationship with other communities. The basis of ecological community relationships on the earth starts from plants. The first utility of direct solar energy and all physical transformation made by solar energy, such as rain, hail, creeks and crevices are captured by the plants and released into the ecosystem for which other organisms compete. Policies intervene in the human attitude on the natural ecosystem and decide how much could be taken away from the resource production system, how the production can be maximized, and how the needs can be served. However, there is a time lag between capturing the energy by plants and growing them to harvestable size.

Sustainability of a production system depends on the matching of the time lag and desire of the policy plan. Most resource policy formulation and evaluation place an emphasis on the relative abundance of resources, either in the form of area or in the form of volume distribution (Castle, 1982); however, seldom is this seen with reference to adequacy of other factors like land, air, water, and other consumptive resources. The nonrenewable resources like rocks and minerals are mainly used by human being. Collection, processing, and utilization of those resources cause displacement and produce a load on the environment. Relevant policies are employed for regulating the volume of use and controlling the distribution of such resources. The characteristic of policy evaluation for sustainability assessment is thus to relate the consequences of use-related operations on the resource systems and their environmental components.

3.3 SOCIAL SYSTEM

The discourses of natural resource operation of developing countries, if we consider the case of forest resources, are mainly concentrated on the causes and background of resource exploitation (deforestation) and the response of specific actors like traders, multinationals, and governments. Governments' unresponsiveness to local and international calls for stopping natural resource destruction usually tries to seek an excuse either by showing the social importance of economic growth and/or to meet the demands of increased population. But some authors (Berghäll and Konvitz, 1997; Kolk, 1996) considered that social changes induced by the governance also have a bearing on the implementation of resource policies. For example, repeated changes in the governance weaken the institutional development due to which a substantial quantity of resource leakage may happen (Bautista, 1990) induced by corruption, nepotism, and bribery. Although there may be many sociopolitical reasons for such changes, some of those changes could have resulted from "cold war," global influence, and/or modernization, which could be seen as postcolonial competition for global supremacy.

Since 1980s, environmental concern for natural resources such as rain forest destruction has been debated strongly in such a way that, irrespective of the state ownership, international communities started to establish a global legitimacy on resources. However, claims of the international communities (hereafter internationalization) on natural resources (e.g., forest conservation) are different in the sense of interest for which they were claimed. On one hand, the claims can be related to undisclosed business interests, and on the other hand, they can be for common reasons. The mood of internationalization usually varies on interest and types of resources and their distribution. In relation to natural resources usually policies assert the rights and legislations, whereas the internationalization issues are guided by arguments and persuasions. The policies are often practiced through depriving people, whereas the legitimate (nonbusiness) internationalization modifies policy practices through participation of people. The policies are mostly autocratic, whereas the internationalizations for commons are democratic. There are clear transitions between policy prioritization and claims of internationalization. Thus, the characteristics of sustainability assessment lie in clarifying social transition for encouraging spontaneous global participation in environmental sustainability.

3.4 GLOBAL SYSTEM

It is understandable that presently the social system is not separable from global system. It is also highlighted in the discussion that in addition to cost–benefit analysis, policy evaluation involves assessment of social parameters. Policy evaluation is thus likely to invite factors from movement of globalization for the coronation of social contemplation. However, the problem is that many authors perceive globalization as a driver of an unsustainable situation. Some scholars (Hirsch and Warren, 1998) sense the smell of colonization in the present global system often calling it neocolonialism. Nevertheless, the concept may have originated in the development literature out of frustrating poverty and lack of sustainability in developing countries. In effect, modernization has been accused of being unable to deliver the goods and services and of an inability to eradicate the poverty. Moreover, modernity made such poverty increasingly associated with ecological degradation as Durning (1990) says:

> . . . poverty's profile has become increasingly environmental.

Under these circumstances, it is hard to see if resources (forests) are used up in search of modernity how a country would contribute to social well-being in a sustainable fashion. Indeed, large-scale resource depletion (deforestation) would tend to reinforce social and economic inequality. For example, Redclift (1991) says:

> forest conflicts are still, after all fundamentally a question of sustainable livelihoods in the face of existing political and economic situation. Inevitably they carry important implication for the way power is understood between groups of people as well for the environment itself.

Therefore, consideration of global initiatives justifies resource (forest) conservation over resource exploitation for economic growth, thus playing a role in changing and improving policies by influencing developmental and economic efforts of a country toward the sustainability thinking of resource management. The global effort for environmental conservation merits even more credit. The momentum of the environmental movement in global initiatives can be grouped into the following actions:

1. national environmental awareness,
2. popular movements, and
3. international environmental negotiation.

The nature of actions and their influence depend on the size and importance of resource. The influence of international negotiation may be predominant in resourceful areas like Amazon and South-east Asia, whereas national environmental awareness is likely to be more influential in the policy changes of less important areas like Bangladesh due to lack of resource abundance. Thus, responses to sustainability issues of natural resources depend, to a certain degree, on the relative abundance and importance of the resources (Castle, 1982).

Despite all these facts and practices in developing countries, the general impression is that policy proponents of the country want sustainability, but unsustainability is mainly caused by resource users. Most policy measures and intervention thus are oriented to control the resource users, not the resource managers. Despite the growing emphasis on policy evaluation, in most developing countries the focus of evaluation has continued to be narrow. In some cases, only financial and physical achievements of a policy are attended. The relationship and coordination between two decisions, either under same policy or of different policies, is seldom considered. Thus, the arguments of policy evaluation should be explained through different approaches.

3.5 TARGET ACHIEVEMENT

One of the sustainability characteristics of policy evaluation is to investigate the functional anomaly of a resource system influenced by the policy measures on a continuum basis and bringing correction to the policy for achieving the goal. Clarke (1995) has mentioned that one of the great scientific dilemmas of the present time is to determine the relative significance of natural and human factors upon environmental changes, because similar physical results can occur from different physical and human processes. The similarity of end results from different processes or factors may occur on the basis of "principle of equifinality" (Goudie, 1984), but evaluation of such end results are difficult for a variety of reasons such as the complexity of natural system and their nonlinear response to changes, imperfection of knowledge, and the possibility of unforeseen extreme events. To avoid the complexity, the policy evaluation may be disintegrated into the following points.

3.5.1 Detection of Changes

Understanding and prediction of environmental attributes and the abundance of resources are important for a better comprehension and/or for taking action in favor of sustainability. We can categorize here six areas, changes in which would bring changes in sustainability. They are:

1. resource growth,
2. climate change,
3. environmental health,
4. changes in governance,
5. future challenges and opportunities, and
6. anthropogenic changes.

Policy processes embracing those issues need to be evaluated continuously to detect the changes, where they are originating from, and for determining which appropriate actions need to be taken. Changes in any of the resource components or factors may bring a series of changes in the environment, some of which may proceed very slowly over a large span of time and thus may be very difficult to identify at the initial stage. Comparative study on past policy actions and their present impacts may help in identifying such factors of changes, eventually that will help in assessing target achievement.

Because policy takes a long time to complete the cycle, over time there could be changes and differences in between the policy in writing and policy in practice, policy experience and policy expertise, judgment value and actual values, policy pragmatics and policy contingencies, social habits and social tradition, and skills required and capabilities to cater. Policy evaluation should assist in detecting such changes as well to ensure sustainability. It suggests that policy evaluation based on contrasting evidence of changes will be more likely to enrich policy and hence the sustainability practice if:

- The changes are detected before it goes beyond the political and institutional limits, creates pressures/concerns on policy makers, fits within their expectation, and resonates with their assumptions.

- The evidence is credible and convincing, provides practical solutions to pressing policy problems, and is packaged to attract the interest of policy makers.

In brief, by making more informed and strategic choices, evaluation for detection of changes can maximize the chances that will influence policy sustainability.

3.5.2 Determining Operation Scale

Limited activity on resource environment ensures the sustainability of resources. For example, a forest resource system comprises of a few interrelated subsystems—some of them are productive to the forest resources and others are exhaustive. If exhaustive processes exceed the productive process, unsustainability may occur. The purpose of policy evaluation is to see the dynamics of components/factors on a continuous basis to create appropriate human intervention so that total exhaustive process does not exceed the total net productive process.

Success of sustainability remains in proactive operations that must be supported by:

1. effective policy environment,
2. powerful policy messages from a supportive political environment,
3. networks, hubs, and partnerships that build coalitions to work effectively with all stakeholders, and
4. long-term programs that pull all of these together.

The policy operators therefore should have clear intent and should equip themselves with skills to uphold balancing actions. This balance can best be defined as the "critical ecological limit." Usually, policy determines operational scale on resources for maintaining the resource balance. Often policies need to introduce substitutes or scientific improvisation of use of goods for balancing the operation scale. Sometimes productive process can also be induced by scientific innovation such as genetic improvement. A sustainable policy needs to be skilful on those issues to keep the balancing action of resource operations by considering the circumstances of other resources.

3.5.3 Harmonizing Operation Sequence

Operation sequence is also an important factor for balancing the critical threshold of a resource system. Even if a small scale of action is undertaken very frequently on a particular area, the resource ecosystem will transform from a close cyclic system to an open system. Though opening of a system gives a higher yield for a short-term basis, on a long-term scale, equivalent amounts of human effort need to be given as an input to obtain the productivity on a sustainable basis (like production from agriculture and plantation). Getting productivity from an open system is equivalent to stretching and squeezing the resource productivity, which cannot be continued for an indefinite period if a certain threshold is exceeded. Therefore, one of the objectives of resource policy is to limit, not only the scale of this type of human operation, but also the sequence of action. The support for effective sequence comes from policy making based on scientific evidence that reinforce operations through:

1. strengthening management capacity,
2. carrying out operations in harmony with human development, and
3. supporting the stakeholders in linking their interest with policy processes.

Thus an extended target of policy evaluation explains the sequence of human action that assists sustainable resource utilization. Under the circumstances, the policy evaluation focuses on management actions pertained to planning and producing policy communication such as policy briefs, research briefs, and stories of change. These eventually help managers and stakeholders to synthesize the outcome of an operation with policy intention. These are then to be communicated to policy makers for corrective action and/or incentive design. Information and communication technology used in operations also indicate a way for managers to better collaborate with stakeholders in sharing sequence of operations.

3.6 ACCOMMODATING TRADITION AND CULTURE

Human action and cooperation for sharing resources among themselves and with other living beings depends on many intersocial and intrasocial factors like culture, tradition, and economy. These variables involve various things ranging from the food habit to burial tradition. The differences in the tradition, economy, and culture may result in a different impressions on the practices of resource use. In some societies, people worship trees as symbol of God. In some other societies using Nontimber Forest Products (hereafter NTFP) is one of the most acceptable traditions of the country people (Byron and Perez, 1996). They use the NTFPs for house implements, traditional treatments, food ingredient, festival article, and many other purposes. Improved marketing system and valuation of these NTFPs may increase the value of forest resources so that they become far higher than the price obtained when they are destroyed (Panayotou and Ashton, 1992; Stiles, 1994). These factors may also increase the acceptability of policy to local people. Thus, a consideration of culture and tradition in sustainability assessment is important.

3.7 SELECTION OF INSTRUMENT

Selection of policy instrument is one of the powerful means for policy success and for controlling human influence on resources. Human needs are very diverse, particularly in the modern society, for which intersocietal cooperation is necessary in the form of trade or exchange of resources in primary or processed form. The trade and exchange of resources can very easily extend and prolong the jurisdiction of the outlet chain of the closed resource system, but for which the whole systems may become unsustainable if not controlled. In a modern society, often selecting traditional instruments may cause policy failure. Generally, the policy target and community obligation and commitment to the policy need to be evaluated before selecting an instrument for policy instrumentation. Usually, various policy instruments are applied for controlling the scale of resource flow applied together to mitigate ups and downs in policy. Selection of multiple instruments are necessary if policy implementation involves several levels of government or requires assistance from a community network. Governance system across scales and levels may be prone to problems of policy implementation. Policy evaluation is expected to provide explanation of how the selected instrument would help in reducing some of these problems to achieve the sustainability target. Such evaluation is also expected to reveal some of the insights as to whether there was any shortcoming in community network consultation during policy formation. Therefore, policy evaluation needs to look into the issues necessary for determining the appropriate instrument and its level of implication for a controlled resource flow.

3.8 INTEGRATION OF DECISION SYSTEM

The implementation of a policy instrument for sustainability of the resource system is mainly affected by over exploitation from the enhanced demand of increased population (Anon, 1999) and/or lack of planning in undertaking measures attributed to the desire of social development (Ostrom, 1999). Therefore, integrating the decision system related to development issues and exploitation issues constitutes a key factor for sustainability of the environment. There could be various ways by which the integration is possible, e.g., increasing economic expertise in environment ministries and/or increasing environment expertise in economic ministry, cost−benefit analysis, adjustment of the system of national accounts could integrate the decisions of economic development and that of environmental sustainability. Though the fields are different, there may be things common in both the fields through which integration issues may be addressed. For example, the factor "consumer" seems to be common both in economic as well as environmental evaluation. Though the term "producer" is common in the economics, in the field of the environment, economic producers can be treated as the consumer of the resources and may convert the resources to a more usable form for the use of direct consumers. Therefore, what are "producer" in the economic climate are direct consumer as well as inducer of the consumption in the environmental arena. While the role of the producers and the consumers and their number, are important for estimating the progress of economic development, evaluation of forest land use policy may consider their role in terms of resource use.

3.9 RESPONDING TO INTERNATIONAL COOPERATION

Over the past several years, the perception of nature and context of international cooperation was guided by environmental issues. But change of global politics, such as the end of the cold war era, has produced a new world order under which the rationale of cooperation has changed and thus influenced the environment in both positive and negative ways. The roles of the World Bank (WB), International Monetary Fund (IMF), General Agreement on Tariffs and Trade (GATT) influence the resource use through structural adjustment, trade, and economic cooperation. There are many global conventions and negotiations, like emission credit, green labeling, and certification, which could influence forest land use. Thus, policy evaluation should also be targeted to the influence of new perspectives of global cooperation.

Often the vicissitudes of the international timber trade may obscure the rich diversity of political and economic forces of modernity that are transforming tropical forests in the South. For example, in South-east Asia, the world's major sources of hardwood exports since 1950s (Ooi, 1990); the trade issues may obscure the importance of population resettlement and agricultural clearance in the regional forest and change is worth noting. In this situation it is important to look at where population pressure and inequitable land holdings, security considerations, urbanization, and the general exigencies of "development" are a central part of resource transformation. To understand changes of resource use in this arena is to address broader political, economic, and ecological questions concerning the inter-relationship of societies and the natural resources at the local, national, and global levels.

From the discussion in this chapter, it is understandable that characterizing aspects of policy sustainability is important to determine the jurisdiction of policy and thereby the extent of policy sustainability assessment required for attaining the policy objectives.

CHAPTER 4

Considerations of Sustainability Assessment

4.1 INTRODUCTION

Considerations for the sustainability of resource and environmental policies depend mainly on perceptions of crises in resource and environmental, and human conditions that prevail in the society. Boyden (1987), Harrison (1992), Meadows et al. (1992), Myers (1993), Brown (1995), and many other literatures have presented such considerations that help in identifying the requirements of policy evaluation. The most important need for facing the present day challenges of sustainability of policies is to establish a firm connection among economic, social, and environmental systems which, Dovers (1996) suggested, can be ensured through a reform in the nature of policy and institutions. The reforms for sustainable policy usually address:

1. A strengthening of social capacity.

Sustainability Assessment. DOI: http://dx.doi.org/10.1016/B978-0-12-407196-4.00004-0

2. The ability of society for carrying a human development program.
3. A linking of society with policy process.

Sustainability of policy in that respect depends on society and policy considerations to address the issues to establish intra- and intergenerational carriers involving with a transfer of social attitudes within the complex of socioeconomic and environmental systems. Brandtland Commission's definition (WCED, 1987) of sustainable development has touched the similar issues comprehensively through the phrase "meeting the needs of present generation without sparing the needs of future generation." If the generational equity is taken as the property of a sustainable system, exploring the basis of origination and establishment of such equity based on the policy may be taken as the activity of policy evaluation aimed at enhancing the sustainability. Dovers (1996) has called it "metapolicy setting;" the setting may only be achieved by a rigorous evaluation for justifying the issues of resource and environmental sustainability as a coherent policy field. The coherence of the resource system and environmental system in a policy depends how the system peculiarities and system components perform in policy practice. The following sections highlight some of the considerations essential for sustainability assessment of peculiarities and system components of resource and environmental policies.

4.2 SOCIOECONOMIC CONSIDERATION

The emergence and the principles of sustainability discussed in Chapters 2 and 3 reveal that the elements of policy climate in a society are important for policy evaluation. A detailed cognition of sustainability issues in developing countries is often available in power relations of the society extending from moneylender to money borrower, farm landlord to farmer, politicians to public, even in some cases government officials to ordinary citizens. The practice of unequal power relations is often linked to conflicts over access to, and the use of, diverse environmental resources (Bryant, 1998). However, the relationship (behind the screen) has a long legacy in policy perspective (Harrison, 1997), particularly in administrative intervention. For example, the step taken by the people in colonial administrative power through changing the status of communal forest land to state-run territories under a different level of legal system, in many cases has marginalized the shifting cultivators, farmers, and poor in the society (Bryant, 1997; Guha, 1989; Peluso, 1992). Even now similar quasi-colonial intervention is not over, but often gets decorated under a fascinating term, "globalization." Therefore, the dichotomy of relationship between colonials and colonized in the past and that of between developed and developing, poor and rich in the present societies is important for policy sustainability. This section aims to discuss the nature and elements of poverty, resource-richness, economy, and institutional development in general to support the policy evaluation.

4.2.1 Nature of Poverty

Poverty is considered as one of the important reasons of resource and environmental degradation in the developing world. Poverty is also considered as one of the indicators of economic evaluation of policy. But in case of environmental evaluation, poverty could occupy many dimensions. Peluso et al. (1994) differentiated poverty according to its characteristics in space, in persistence, and in its identity. Poverty is usually common in Natural Resources Dependent Areas (NRDA). By NRDA, the authors meant the places where natural resources either account for a substantial part of the local economy or attract population. The NRDAs in their turn can be classified into three categories:

1. Extractive NRDAs are places where the local economy is based on renewable or nonrenewable resources.
2. Nonconsumptive NRDAs have local economies based on nonconsumable natural resources like tourism.
3. Backdrop NRDAs have natural resources serving as an esthetic backdrop, luring residents whose income usually originates elsewhere.

The classifications of NRDAs are not mutually exclusive; all of them may occur in a single economy, distributing themselves in the space by conservation of significant natural resources or in urban areas. Therefore, comments or conclusions about poverty and sustainability may not always be applicable to all NRDA areas; rather they indicate different opportunities.

If overcoming poverty is taken as one step ahead to sustainability, it is important to evaluate the dynamic nature of poverty. According to the dynamic proposition of poverty and its persistence, poverty can also be categorized into three:

1. Poverty is a state through which households and individuals pass during certain life-cycle events but at different point of time individuals or whole households may move in or out of poverty.
2. Poverty may be tidal, the rate of poverty may rise or fall, but certain individuals or households remain poor regardless of fluctuation.
3. Poverty may recur potentially for a whole community or may strike a community in such a way that it is physically, economically, or socially devastating.

Whole communities and even whole regions may be impoverished as a result of their political, economic, or geographic locations, which are permanent in nature. Often poverty may be stricken by temporary political victimization. The aspects of poverty need to be considered as emerging issues for sustainable policy. Poverty often causes the loss of social cohesiveness and thereby sustainability approached through social policy and planning does not work.

Poverty also varies according to the identity and status of a person, e.g., Peluso (1992) contended that poverty is more prevalent among females, children, and old people. Social cohesiveness depends on where the poverty really strikes. Poverty may strike depending on the resource pricing. For example, if pricing is such that it encourages non-consumption, the resource traders will be the losers, and the resource management will be difficult if the authorities cannot check the illegal trade. Poverty may be structural as well, e.g., saw milling, if moved to built-up sites, other than the forest sites, would affect the economy as well as physical environment. Structural poverty, if identified, can be improved by appropriate policy measures, but if more than one type of poverty exists, then it becomes very difficult to maintain the sustainability principles. Under those circumstances, sustainability issues could be classified according as their relationship to NRDA type poverty or the dynamic nature of poverty and can be dealt with as one type of problem at a time aiming at the total policy build-up over a time frame and/or the resource frame.

4.2.2 Nature of Resource Availability

Resource availability is an important factor of social poverty and thus of policy sustainability. Even the term environmental pollution is intricately related to resources, such as air, water, and soil pollution. Thus, although presently more abstractive classifications are introduced, e.g. cultural pollution, yet most environmental legislation delineates the protection of resources from pollution. Those classifications have facilitated the formation of different legislation and have been described as "compartmental division" of legislation (Hajer, 1995).

Compartmentalization of legislation was influenced by some elements of policy sustainability such as the nature of resources, market-oriented investment policies, and commitment to sustained supply. Gaventa (1980), Marchak (1983), and Peluso et al. (1994) recognized that the market-oriented investment policies of overseas capital extract resources destructively. Multinational timber companies, for example, enter a region and exhaust natural resources without concern for the region or its people, but the markets remain far away from the country. This is comparable to the practices of colonial times. The long-term costs of such extraction are thus borne by local people. Therefore, the controllers of the resources and their roles are also important element for policy evaluation.

4.2.3 Nature of Economy

From the above paragraphs, it is understandable that the standing of sustainability is different in different countries depending on the status of the economy. Some nations have to bear the burden of a large population, some countries seek economic growth, and some others have plenty. The most important aspect of policy for sustainability is to identify the concern of unsustainability. For example, if NRDAs are a backdrop type of which the economy is not based on local resources, according to Nord (1994), sustainability will largely depend on the vicissitudes of the national economy. In this process, urban poor flee to rural areas hoping to find more affordable housing, more respectable surroundings, and an escape from urban violence. Similarly, in nonconsumptive NRDAs, the fortune of the tourist industry/ies will rise and fall with social order and fascination of nature. The destructive forces of natural beauty and resources, such as forest fires, become a part of prescription for sustainability planning.

However, natural resource dependence is not a prior cause of poverty and hence sustainability. There are some other causes, which may bring a sustainability risk to NRDAs. These causes are important for policy evaluation and can be outlined as follows:

1. Centralized economic structure—if dominated by a large single company or government enterprise whose major interest or activity is extracting and selling a single unprocessed raw material (Bunker, 1984).
2. Technological inability—processing resources to a poor-quality end product, fetching less market value, and excessive wastage (Blaikie, 1985; Freudenburg, 1992). These may be added to inefficient resource extraction, lack of local investment, and lack of economic multiplication.
3. Concentration of ownership and control—a likelihood of absentee ownership due to a centralization process and sector dominance by large firms (Freudenburg, 1992; Marchak, 1983).

Thus, policy climate inevitably becomes linked with control of resources, such as resource dependence, resource use, resource waste, and nature of capital (e.g., external or internal). Sometimes structural processes related to natural resource dependence may cause both poverty and environmental degradation (Blaikie, 1985). Moreover, weakness in structural processes retards the institutional development. Thus, the policy climate gets polluted by weak implementation leading to erosion of human behavior in terms of nepotism and bribery. Hype (1994) observed that environmental erosion and social erosion are so closely associated that one cannot be studied without the other.

Peluso et al. (1994) argued that such behavior in government agencies and structural processes may degrade public land resources affecting local livelihoods from alternative land uses and sustained job opportunities. A government agency may act like outside capital but justifies its action by its self-proclaimed representation of the greatest good for the greater number. Therefore, the end result of sustainability depends on the nature of capital and its management.

4.2.4 Nature of Capital
The nature of capital determines the scale of operation in the resource sector and thus influences the players' role in policy climate. Advanced capitalism signifies concentration of capitals, vertically integrated, increasingly capital intensive, oligopolistic industries, and segmented labor markets and results in withdrawal of local control over economic

process and decisions (Markusen, 1987). The capital intensive firms can extract resource globally. The processing efficiency of capital intensive machines allows the extraction of unprecedented volumes of raw materials (Bunker, 1984); thus, the influence of capital is not only spatial but also on scale of operation accentuating unsustainable practices. Due to competition with technologically efficient and capital intensive processes, NRDAs may experience changes with widespread mill closures and loss of jobs (Marchak, 1983). The state thus needed to protect those mills through subsidies. At an extreme situation, the NRDAs could be transformed into net primary produce exporters, fetching little earnings in return with very little value addition. Thereby, they need to export more resources to meet the requirement of desired growth. As a result, growth in direct NRDAs occupies more space than that of sustainability prescriptions in the policy climate.

The transformation of advanced capitalized productions may cause a shift from resource use to alternative resource use with short-term economic benefits. Thus, the nature of capital determines the stability of institutions and organizations for supporting sustainability. Primary industries in a resource sector bear a disproportionately high part of the business instability partly because of the nature of resources and the mode of their extraction (Bunker, 1984; Freudenburg, 1992). Therefore, adjustment in the roles and modes of institutions and organizations can be a significant factor for sustainable policy.

4.2.5 Nature of Institutions
The displayed decision of a policy depends on the nature of institution whether it is bureaucratic or democratic. Non-openness in decision-making, delay, and red tape in bureaucratic control reduce public participation, delay project implementation and hence impede the sustainability. Sometimes the institutions and organizations of developing countries cannot work completely due to political barrier, dishonesty of staff, and systemic error. The institutions of such societies take risks on environment to get wealth. On the other hand, a society could enjoy the environment but without a guarantee of monetary wealth. Thus, a society that becomes exposed to risk is a construction of the nature of the institution or the ability and wealth of that society. Beck (1994) has termed such society as a risk society. Thus, the thinking of a risk society is a discourse of self-confrontation, therefore, it is less likely to be sustainable.

In such societies, claims of scientific truth may bring the experts as authoritarian forces that may clash with politics (e.g., valuation for environment) and can hinder progress. However, for political reasons social authorities (actors) may not adopt the full truth of scientific claim. Thus, imbalance in self-confrontation may also risk a society. Under that situation, the institutional role may lead to a negotiation appropriate for the society. Therefore, the nature of institution, how it has developed, and how it functions do matter as to how the situation would be negotiated, and thus how the risk of unsustainability could be reduced.

Strong policy commitment is also important for bridging the ideas, because sustainability is not a single discourse rather it is a collection of specific ideas. Concepts and categorizations of sustainability discourses that are produced, reproduced, and transformed in a particular set of actions through which a set of ideas can be given an institutional or organizational reality (Hajer, 1995). Thus, sustainability discourses can be categorized as the set of actions credible to actors or usage of culture or role of institutions. In practice, institutional rearrangements are seen as the precondition of the initiation of the sustainability process (Dovers, 1995) and the role of actors and usage of culture contribute to the realization of sustainability discourses through organizing peoples' perception, policy process, and political procedures. In sustainability assessment of policy, therefore, it is important to investigate how institutions are made to operate in line with the conception of sustainability.

One of the ways to initiating institutional procedure is through subject positioning and structural positioning of the discourse, which may create institutional machinery for investigating sustainability. This idea is also supported by Blaug's (1992) statement that fragmentation or categorization is a more general characteristic of investigation practices. Walsh et al. (1999) advocate two sets of social variables as critical for investigating institutional ability of sustainability:

1. Those which measure the human capability to alter the environment.
2. Those which measure the social and institutional constraints outside the control of individuals, household, and even the state.

Both human capability and institutional constraints can be linked to the policy processes that the society experiences. Whether it is democratic process or autocratic abuse, sustainability cannot be expected if the inability and constraints are not removed. Although international facilitation is there to help in removing some of the constraints, it may not always be possible because economic globalization is moving faster than the ecological globalization (Agarwal, 1998). Here also the institutional role is important.

Economic globalization works through obligatory financial rules set out by GATT or the United Nations Conference on Trade and Development (UNCTAD) and is implemented through obligatory custom and excise departments of individual countries. But ecological globalization works through voluntary international conventions like Montreal, Oslo, Rio, or Kyoto conventions and only occasionally by mandatory international environmental rules, like those of Convention on International Trade on Endangered Species (CITES). Thus, the institutions in developing countries extending their flow chain with globalization become more vulnerable to financial regulations than to the ecological regulation and become engulfed by the competitive economic situations before taking any benign initiative to ecological adaptation.

Moreover, the problems are exacerbated under the weak policy mechanism of developing countries. Political and commercial actors weaken it further when institutional arrangement and incentives faced by such actors are flawed. Often the phenomenon of government failure in the developing countries, especially when the cost of government action outweighs the benefit of intervention, causes an indirect failure of policy implementation. Additionally, political actors (politicians, bureaucrats, and voters) in developing countries may operate under undesirable incentives which alleviate inefficiency in the institutional mechanism. Special interest, lobbying, information problems, short political time horizons, instability, and political expediency guarantee failure of institutions and other policy mechanisms. The general impacts of some of these components reveal that the role and nature of institutions depend on the characteristics of the society and social behavior in the policy climate developed by the administrative influence on resources and culture of the society.

4.3 CONSIDERATION OF SYSTEM PECULIARITIES

In almost all the environmental cycles, the role of natural resources is prime and foremost which has made its distinctive position for "metapolicy setting" and precedence for mobilizing sustainability drive across the globe. However, being a natural entity, the resource system has typical peculiarities that need to be emphasized in policy evaluation. For example, when the decisions on forest land use are made, appropriate "metapolicy settings" for environmental sustainability may obscurely add some additional peculiarities (nonsystemic) to the natural resource system. Evaluation of resource policy needs to consider those peculiarities. Some of those peculiarities are presented in the following sections.

4.3.1 Temporal Scale

Natural system function in a resource system is very slow and occurs on variable timescale. Most resource systems (e.g., forest) take a very long time to reconstruct the cycle, far greater than a political cycle or electoral cycle, which is highly influential in determining policy issues, but of which the responsibility of wrong policy goes to others. That is, the functional period of "political ecosystem" does not match with that of the resource ecosystem. This causes problems when future environmental issues of natural resources are predicted and expressed without considering the margins of "political ecology." Often margins of temporal scale are addressed by short- and long-term goals. Our understanding goes that sustainability concept is associated with long-term objectives; however, the short-term achievements cannot be neglected in sustainability evaluation. Long-term goals of a resource system usually take several political cycles to achieve. Moreover, the fundamentals of long-term objectives mostly depend on achievements of several short-term objectives. Therefore, for a sustainability evaluation it is important to see how and whether the social system can assimilate those short-term goals within the limit of political and economic aberration.

4.3.2 Spatial Scale

The extent of resources and their distribution matters a great deal for considering short- and long-term perspectives of policy goals. Often substitution and complementarities of resources come into play to mitigate the shortcomings in the spatial scale of resources. Spatial scale of resource system not only means the geographical spread of the policy influences but also means the influences across other policies. However, environmental problems like air pollution and river salinity can spread horizontally as well as vertically across the space irrespective of political or horizontal boundary. Thus, when the temporal scale of a policy denotes an intrapolicy arrangement of time horizon, spatial scale of a policy demands a concern of interpolicy coordination within and beyond the states. This may be between resource and environmental policies or other policies like trade or tariff policy. Thus, spatial dimension of policy evaluation of a country merits an integrated scenario of intra-/interpolicy linkages beyond the jurisdiction.

4.3.3 Connectivity and Complexity
Analyzing intra- or interpolicy linkages is not simple. The complexity comes from differences in connectivity of different policy goals and social desires and their dynamic nature. Additional arrangements may be necessary for evaluating and explaining the interconnection of policy goals and problems with other social and economic policies. Usually, tools like environment, development, economy, poverty alleviation and social well-being are brought into the scene to explain the connectivity. For resource and environmental policies, issues like soil erosion and siltation, energy and transport system, resource harvesting and greenhouse gas emission can be mentioned as examples. Thus, sustainability evaluation of a policy has to be visualized from all angles of social, environmental, and economic implications. For example, the result of a good education policy is expected to bring good environmental awareness across the community as well as preferred employability of graduates. Evaluation of such linkages is undeniable.

4.3.4 Accumulation
Most of the impacts of resource use on environment are not discrete but cumulative. Such cumulating events may be horizontal as well as vertical. For example, if a person clears virgin land and gets higher productivity, other persons of the society will also tend to do so, which is an example of horizontal accumulation across the society. However, if such practices go on, extinction of insects or loss of nutrients from the land will be progressively multiplied, which is an example of vertical accumulation. Also, accumulation may be catalytic when one factor may induce the accumulation of other factor of resource use. For example, accentuation of soil erosion will accentuate the damage to river ecosystem. These accumulations happen over time and extend over space; thus, evaluation of cumulative impacts of policies can be assumed as future condition of present resource use reflecting a compounding impact of past activities. This is not similar to internal rate

of return (IRR) or cost—benefit (C/B) ratio used for economic analysis but more to see whether such accumulation will override the social tolerance in a future perspective.

4.3.5 Nonmarketability

In the case of resource and environmental benefits, some goods and services are not marketable and cannot be termed in the economic parameter. Particularly, in the case of natural resources, people have to go to the resource site to enjoy services. Even though valuation of some environmental output of resources may be possible by opportunity cost or by past value; they are subjective. Moreover, such values cannot be calculated back to the history by normal IRR method. In many cases, qualitative explanation have been given to put the case forward. However, justifying policy formulation on the basis of qualitative explanation is difficult. Thus, sustainability of resource and environmental policies require significant effort to reveal qualitative considerations into monetary connotation. Sustainability assessment assists in this regard to overcome difficulties in the nonmarket nature of a resource and environment relations.

4.3.6 Moral and Ethical Considerations

The resolution of all policy problems entails moral and ethical issues. However, environmental sustainability problems raise moral questions that are new, noble, and include issues of intergenerational and nonhuman species rights (ecological). However, the relative importance of moral and ethical issues of a policy depend on the ruling authority. The moral or ethical issues of ruler represented by the society (democratic) and ruler not represented by the society (autocratic or colonial) may not be necessarily the same. Similarly, morale and sympathy of a local ruling authority could be different from the central ruling authority. However, a prolonged existence of certain type of morale (tradition) may influence the attitude of people permanently. Therefore, sustainability evaluation of resource and environmental policies may need to look at the adjustments and evolution of morales and values incorporated within the society; whether they are political, societal, or traditional.

Having mentioned the peculiarities of resources, it is important to remember that the debates of resource use and issues relating to resource and environmental sustainability usually come together.

Particularly, the influences of forest land use on the environment make this inevitable. However, a policy involves issues and motives for social benefits. Thus, evaluation of a resource (forest) policy significantly involves the issues of resources as well as their relationship to the ecological, environmental, and economic systems. Although the term "environmental sustainability" embraces the sustainability issues of all the contingent branches of environment, the interest of this book is to investigate how environmental sustainability can be oriented in resource policy.

4.4 CONSIDERATION OF COMPONENT PECULIARITIES

The nature and importance of system peculiarities may be different for different components of policies. The sequential arrangement of policy evaluation largely depends on the evaluation of components. At the same time, components of a policy system may have spatial variation. If a policy is evaluated without emphasizing components, then it is likely that the spatial variations would be overlooked. Spatial differentiation involves the variation in the importance of ethical values of resources depending on the socioeconomic condition of different places. The sequential variation depends on the component of the policy system. Therefore, policy evaluation needs to look at the system components and ethical components in their sequential and spatial distribution. Figure 4.1 shows the dimension of social, physical, and environmental components related to sustainability assessment of resource and environmental policy in their sequential and spatial relationships.

Figure 4.1 shows that sustainability issues can be considered under three broad definitions of environment: physical environment, geographical environment, and ecological environment. These environmental connotations are so closely interrelated that discussion on any one of them in addressing the sustainability issues of environment will touch the other two. Indeed, sustainability of one separated component is not possible; it includes all the components of a system (Fox, 1996). On the one hand, sustainability of an environmental system contributes to the resource sustainability; on the other hand, resources contribute toward the sustainability of environment. Their mutual relationship of sustainability as maintained through the interaction of their component network in space, as shown in Fig. 4.1, reveals that policy and sustainability are the bridges. If the network evaluation of components

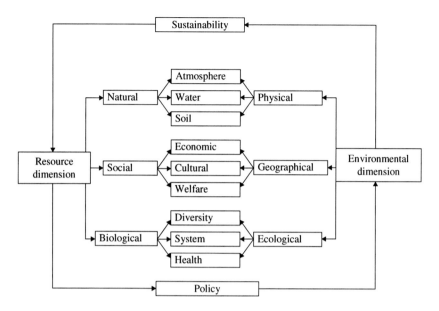

Fig. 4.1 Considerations of policy sustainability involving resource and environmental dimension.

is found congenial, one can say that the sustainability of policy is achievable. In practice, policies are designed to correct the mismatch among network components. Although sustainability assessment of a policy admits the evaluation of all three aspects of the relationship between resource and environment as decreed through Fig. 4.1, within the limited scope of this book we will look into the social aspects only.

The status of societies and nations differs from each other in various ways. Although some societies could have some regional similarity in culture, tradition, economy, resources, and living condition, they are largely affected by geoenvironment in which they live (Ojima et al., 1994). There might be regional differences in the technological capacity and wealth that can affect the policy climate and resource use pattern. Such affects may become prominent through status of trade, negotiation, and conflicts of interest. Therefore, the elements of changes are embedded in the socioeconomic and environmental build-up of the society, prioritization of need-base preference of the people, the dynamics of which often dictate how a resource is used and will be used within a region. Thereby, evaluation of resource policy requires understanding of environmental, economic, political, demographic, and other social conditions and how their influences have been considered in the policy measures.

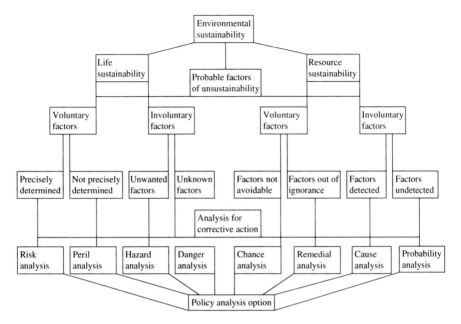

Fig. 4.2 Forms and factors of sustainability considerations.

Policy works in society regulating the relationship of society to the resources and the environment, whereas sustenance of resources and environment are regulated in their own space by nature and timescale. Thus, the sustainability of a policy is a balance of behavior of society to the ability of natural sustenance of resources and environment. If there are imbalances, there could be some hazard or risk. There could be many factors involved in bringing such imbalances into the limelight of evaluation. Figure 4.2 summarizes the different types of factors and the evaluation pattern required to corresponding types of assessment.

From the interaction of components in Fig. 4.1, environmental sustainability can be conceptualized as life sustainability (e.g., diversity and health) and resource sustainability (e.g., soil and water) of Fig. 4.2. Fig. 4.2 demonstrates that unsustainability in life and resources may occur due to voluntary or involuntary factors. Although one may argue that unsustainability cannot be voluntary at all, but there could be some voluntary factors produced from ignorance or personal interest. However, involuntary factors may have their origin in the past where present system has little control on any environmental catastrophe originated from the factors of the past. Under the circumstances, evaluation of policy steps in and after implementation is more likely to bring accurate result on sustainability. As a result, an inherent component of policy is that it has to be flexible for correction from feedback analysis.

As this book focuses on sustainability assessment of policies of resource systems, the aspects of evaluation will remain limited to within few options. If the factors of unsustainability are voluntary, e.g., for some reasons of personal gain of ruling authorities, policy makers can do little, because it is not the fault of policy itself. Most often, particularly in developing countries, policy makers cannot avoid voluntary factors due to pressure created by people in power. Thus, most policies consider involuntary factors to prescribe a minimum corrective action such as learning from the historic factors, cultural reasons, or probability analysis for undetected factor, e.g., increase of demand. Often policy skill of presenting involuntary factors may create pressure for controlling voluntary factors. Thus, the evaluation of a policy targets certain operational processes to display the interplay of voluntary and involuntary factors in sustainability. Chapter 5 delineates some of the operational processes of resource policy (e.g., forest land use) as a research target for policy evaluation.

Issues of Sustainability Assessment

5.1 INTRODUCTION

There are certain issues on the basis of which it is possible to identify the nature of policy sustainability. Environmental concern can
be considered as the most important of all the issues. Initially, environmental problems were treated as an ad hoc or expost problem
requiring some remedial measures. The understanding of environment as a structural problem was realized toward the end of 1980s,
which resulted in an emergence of environmental discourse called

Sustainability Assessment. DOI: http://dx.doi.org/10.1016/B978-0-12-407196-4.00005-2

"ecological modernization" (Hajer, 1995). Ecological modernization assumed that existing environmental problems can be internalized by political, economic, and social institutions and can care for the sustainability. The concept of "ecological modernization" as introduced in the discourse of environmental sustainability can be outlined as follows:

1. It assumes that environmental degradation is calculable most notably by combining the monetary units and discursive elements from different branches of natural and social sciences of which cost−benefit analysis of environment is one of the examples.
2. It assumes that effective management of environmental problems is possible by collective action; if every individual, firm, and at the larger arena if every country participates, management of sustainable environment is possible.
3. Economic growth and the resolution of ecological problems can be reconciled; that is, the discourse of environmental sustainability follows a utilitarian logic.

Sustainability thus covers a broader realm than that of environmental concerns which utilizes the basis of ecological modernization and observes the following few realms:

1. A shift in environmental policy measures. The old principle of "react and cure" has been replaced by an "anticipate and prevent" discipline.
2. Integration of compartmental division. Division of environmental boundaries through regional pact and/or international convention.
3. Inclusion of deregulation methods replacing the hierarchical legislative system involved in environmental management. The organization or institution itself will be responsible for environmental management. As a result, last few years have witnessed impact assessment, risk analysis, polluters pay principle, cost− benefit analysis, precautionary principles, tradable pollution rights, levy of charges as well as debates on resource rent and emission taxes.
4. Influence of science in policy making. Science, as an evaluator of damaging effects of environmental and social change, provides a basis for an exact input for policy making. For example, the concept of critical load and more benignly the concepts of multiple stress. The concept of multiple stress has transformed the

reductionist scientific view of "critical role" to orientalist idea of "integrated role" (NCEDR, 1998).

5. Replacement of unproductive investment at microeconomic level by anticipatory investment (e.g., filter technology in industries). At the microeconomics level, the shift toward environmental sustainability drives the idea of "environmental protection solely increases the cost" to the concept that "pollution prevention pays back." The inclusion of the concept in the policy has led to promoting low and nonwaste technologies in developed countries and put forward the idea of "multivalued auditing" (success is measured not only in terms of money but also taking energy and usage into account).

6. Conceptualization of nature as "public goods" in the macroeconomic level. Nature used to be treated as "free goods." Nature including forest was utilized as a sink, but under the present concept this cannot be done without paying for it. Therefore, externalization of economic costs of environmental pollution has been reduced and thereby, the concept can put a great emphasis on the conservation of natural resources by stimulating ecological pricing, recycling, reusing, and technological innovations.

7. Change in the characteristics of legislative discourse. Modern approaches assume that environmental sustainability of resources would depend how the price of the environment is included in the resource valuation (OECD, 1985). In this discourse frame, statistical prediction has become the basis for collective liability of policy sustainability.

8. Consideration of participatory practices. This concept seeks a plausible solution between the requirement of resources and minimization of problem. Therefore, it acknowledges the most proliferated and effective actors like NGOs.

These principles and objectives show that the policy climate for environmental sustainability could be pursued through ecological modernization. This could lead to the increased influence of science and decreased influence of legislative and regulatory system in policy climate. Understanding may take the place of compulsion in management. Thus, environmental sustainability possesses some special dimensions in policy climate that are important for considering transitions to policy evaluation. The following section presents a brief review of basic issues for policy evaluation.

5.2 ISSUES RELATED TO SOCIETY

On the basis of social theme, policy climate around resources and environment may be considered as a part of the society created by society for society. In developing societies, policy climate is not coherent but inherent and contesting (Wilson and Bryant, 1997). Within the society, there are environment users as well as environment managers with contradictions among and between themselves for right and tenure of resource use. A policy within those contradictions finds credibility to offer a regulatory action or strategy; but sustainability matters whether policies are acceptable to society and generate trust to the institutions that are put in the charge of regulation or not. This is a form of "discursive closure" for solutions of sustainability problems where variables for policy actions are actors like policy makers, managers, and their critics from people and institutions. Within a society, the role of these actors helps to avoid social inherence and contradiction producing coherence and a holistic approach of sustainability. There are certain ingredients of the society that determine how the coherence will develop in a policy system. The following sections will highlight some of those ingredients.

5.2.1 Social Modernization

Social modernization offers an understanding of changed social perception of particular problems around which actors get an opportunity to act more coherently. A dynamic society deals with a social problem by dissociating it into specific causal components. However, there are associated converse comprehensions of social modernization as well. The concept of "risk society" and "reflexive modernization" can be taken as the basis for the discussion of the role of modern society in environmental sustainability. According to Beck (1992), "risk society is an idea of forceful achievement of modernization which basically follows the historical era of industrial society." While industrial expansion and growth become the motto of developing societies, the equation of risk taken by such societies needs to be resolved.

The difference between the role of risk society toward environmental sustainability and that of traditional society begins from the nature of the comprehension of environmental problems. Indeed, some of the environmental problems were not merely incidental but incremental (e.g., global warming). Wilson and Bryant (1997) claim that the difference in the comprehension of nature and the environment of societies gives the reflection of different time and space. The end result of the environmental observation of nations will differ markedly between the periods as well as among the nations depending on the progress they have achieved on the wheel of modernity. That is why environmental discourse reveals and claims a routine check of social roles on the issues of policy evaluation of that society over that time.

5.2.2 Societal Relationship

The societal relationship induces societies to be coherent. Issues related to trades and transfers of resources mostly depend on societal relationship. The components of a policy dealing with the environmental crises thereby should have their roots in the relationship of society or social organization. Attitude of social organization and social policy toward environmental resources delineates sustainability standards. The attention paid by a society to achieving sustainability is termed here as "social bias." Attitude to social organization is thus responsible for mobilization of biases, which is important to incorporate societal relationship in the evaluation of environmental sustainability of any resource policy. The organizational attitude and the social bias depend on their relationship with history, past traditions, and customs that justify the demonstration of social bias. Such a relationship is also important for global societies. Because resource policy evaluation has a spatial scale, differences in societies in consumption or attitude to resources could have a significant effect on resource sustainability. Therefore, the bridges of state−society relationships are important for policy evaluation.

5.2.3 Radicalization and Convergence

The terms radicalization and convergence are considered here to mean the aloofness and association of state policy with the characteristics of society. While radicalization indicates some kind of peculiarities of policy or society that do not correspond with each other, convergence means adjustment of peculiarities persuaded towards achieving sustainability. Social and economic inequity is a common characteristic of developing society. Thus, if state policy does not work for a poor mass, a large proportion of the community suffers from the impacts of a dreadful policy. As a result, the environmental negligence in the policy, though it seems small, actually multiplies in the social space. Therefore, the key concept of environmental sustainability also emphasizes controlling the marginality of the people to reduce vulnerability. That is where the aim of the state should converge with the aim of society.

For sustainability reasons, the convergence of cultural lifestyle and the environmental component of resource, culture, and relationships of social groups (e.g., minority) cannot be avoided in the resource policy (Colchester, 1993). In many parts of the developing world, forestry has now become a part of wider social practice under the impetus of social forestry or homestead forestry (Carney, 1996; Jokes et al., 1995). These authors recommend that unsustainability issues of forest resource used in those societies need to address the equity in gender differences in terms of economy and power relationship. Thus, along with state–society convergence, intrasocietal convergence is also necessary for policy sustainability.

The impact of differences in power relationship on unsustainability may also extend beyond the society and may prevail in the international arena. Bryant (1998) noted:

> at one level power is reflected as the ability of one actor to control the environment of another. Such control may be inscribed in the environment through land, air and/or water alterations: felled forests, timber plantations, cotton fields, toxic waste dump, mine tailing, barrages over the river, smog from the forest fire and so many.

Thus, in developing countries though the degradation of environmental resources is commonly blamed on the fuel wood gatherer, shifting cultivators, and encroachers, evaluation of existing power relationships and business attitudes among and within the states may shift the blame of deforestation elsewhere. The policies for controlling power, politics, cronyism, trade, and economics may appear as the salient reasons for changing attitudes of poor fuel wood gatherers and shifting cultivators.

5.2.4 Boserupian/Neo-Malthusian Issues

Boserupian and Malthusian concepts characterize the relationship of people and resources, two most important components of society. Instead of taking population or technology as separate factors, Boserupian influence has considered a social characteristic to mean the combined effect of technology and population. According to Boserup (1983), if other methods of obtaining food or change of diet are not possible only technological advancement is enough to meeting the increased sustainably needs. Under the circumstances, intensive pressure is exerted on the land to meet the staple requirement of people. However, neo-Malthusian concept assumes that, using technological advantages, sustainability can be maintained. Although the concepts sound like contradicting, actually they are additive— Boserupian concept is an additive to neo-Malthusian.

However, whatever concept is adopted, there is no doubt that they will produce intensive pressure on land production system that may cause a reduction of soil productivity in the long run and therefore, a permanent solution to sustainability issues will remain remote. People will start to use fragile lands, marginal lands, and slums (FAO, 1997). Thus, what seems to be a technological/social debate in neo-Malthusian and Boserupian concepts, could be turned to economic inequality. Therefore, the social coherence would be hampered, and thus the sufferings of the large poor mass could multiply in terms of environmental degradation.

5.2.5 Social Ignorance

Although social ignorance means many proletariat issues, in relation to policy social ignorance means a construction of policy inducement for hiding facts by removing details of reality away from the discussion and/ or overriding facts by other less relevant information. This happens mainly for the prestige of bureaucrats, to hide the dishonesty of officials and politicians or to direct the societies in a particular way. Negligence to knowledge of traditional people or aboriginals may also lead to a situation similar to social ignorance. Although such construction may be possible for positive action, in most cases in developing societies it happens for negative actions thus influencing policy sustainability in a negative direction. For example, transmigration and deforestation may have a strong correlation, but due to political motives the information may not be supplied in discussion. Therefore, social ignorance is also an influence of actors who are social elements. Policy evaluation should not avoid such a construction of social ignorance.

5.2.6 Social Attitudes

Social attitude may be considered as the expression of a society or social majority toward a particular policy or condition. Social attitude may be treated as a translation of the combined effects of all the social factors, similar to the translation of the Boserupian factor to socioeconomic factors. To a particular condition, such as environmental sustainability

of forest land use, social attitudes indicate society's willingness to participate or to deny the proposition and can be determined by the status of
policy or the resource itself. Thus, in the specific case of environmental
sustainability, social attitudes involve both technical issues of resources
such as: status of forest types, species, productivity, processing and utilization, and socioeconomic issues adhered to the people, their culture,
education, income, and supply–demand variables. Thus, if social attitude is considered, assessment of policy sustainability does not depend
on the resource or environmental system only but also on the sociocultural system.

The relationship between resource and culture may be critical for
social attitude. For example, although many authors blame colonization as responsible for the resource degradation of many developed
countries, they did not mean that all the resources were exhausted during colonial time. However, as the sociocultural development was not
maintained, the liberation of social attitude developed during colonial
times produced an unsustainable situation later—during postcolonial
times.

The discussions in the above few sections show that in developing
countries, where population is very high and socioeconomic condition
is not very progressive, it is reasonable to assume that social issues are
crucial to the sustainability of resource conditions and should be part
of the resource policy. Therefore, an understanding of policy elements
is important for adjusting social elements in sustainability assessment.

5.3 ISSUES RELATED TO POLICY DISCOURSE

From the discussion in the previous sections, it is understandable that
appropriate policy implication is important for organizing the social
issues toward sustainability. New elements, instruments, and strategies
as well as emerging works of supernational organizations (secondary
organizations) are backing the policy implication toward sustainability
(Hajer, 1995). This is an example of "discursive action" equivalent to
actions of "paradigm shift" as conceptualized by Mather (1997).
Although this conventional discursive approach seems different from
the sustainability concept, in fact various practices at microlevel could
influence the way in which sustainability is interpreted to make the
resources manageable and available for the future needs of a progressive

society (OECD, 1985). This point of sustainability is to show the sheer variety of ways in which policy discourse influences the process of social change and how the reproduction of the demise of social structures depends upon the outcome of discursive interaction. The elements of such discourses can be treated as policy strategies. Some of the elements of policy strategies are discussed in the following sections.

5.3.1 Discourses of Story Line

The concept of "story line" was used by Hajer (1995) to mean recurring figures of speeches that dominate public understanding of sustainability, which help in rationalization and naturalization of the existing social order of developed countries. For example, in Britain, a story line can be built up on any sustainability issues addressing the period before and after the World War II. Usually, people in Britain become very much emotional as well as proud over the story line referring World War II. Thus, a story line of historic importance and of national pride often works very well if can be integrated with the policy line. People of Bangladesh are also becoming very emotional on the history of their independence war, which can be used successfully in the policy line. In practice, each nation has some events/issues in their history the discourses of which can be taken as a strategic element for motivating people. Thus, story lines are social events used as elements for sustainable climate policy.

5.3.2 Discourses of Disjunction Maker

Policy-making institutions of a country had their own legitimate ways of denying the institutional dimension of the sustainability challenge. These so called "disjunction makers" constitute essential elements in the policy evaluation. For example, in Britain if scientific evidences constituted an unavoidable hurdle for any attempt of changing environment policies, business corporations may lobby for avoiding an environmental concern to minimize the cost. Such agents may be termed as "disjunction maker." In developing countries, though raising people's awareness may be used as potential force for sustainable policies, inadequate information and evidences of scientific dichotomy may impede the awareness producing an effect similar to a disjunction maker. Thus, disjunction makers are unfavorable social elements normally prevailing within the society but appearing as disjunction makers when publicized repeatedly by actors/players in forums like social media. Therefore, policy evaluation engages in an explanation of such elements because of which a policy may become unsuccessful.

5.3.3 Discourses of Symbolic Politics

The term symbolic politics means the social elements to which most attention of policies is driven. Different policies may have different strategies to take sustainability problems within the policy. For example, in developed countries, if public awareness grows on a certain aspect of the environment, then government may immediately seek to take some policy measures. If successful, the success becomes the political achievement of the government. Therefore, it is expected that the quality of environmental policy will improve day by day. In developing countries like Bangladesh, such things rarely happen, mainly due to lack of funding and sometimes lack of expertise. The political symbols remain oriented around hunger, thus in many instances, environmental issues are politically overlooked. However, improving hunger and poverty can also be symbolic as environmental improvement if done not for a mere political gain but for a sustainable improvement. These examples signify that political symbols should somehow be included in policy elements used for generating sustainable social impact.

5.3.4 Discourses of Sensor Component

The sensor components are social elements that can be used as an important tool for triggering a feel for policy formulation; particularly noneconomic issues like environmental sustainability. Often meetings and conferences may trigger the sense of a policy maker for undertaking a policy or to identify whether a policy target is right or wrong. If policy makers visit the problem area and study it, the problem initiates the concern for formulating policy. However, sensory incidents are different for different societies, their education, and culture. For example, once folklore was favorable, sensor equipment for initiating policies in Bangladesh. People used to go to village markets and schools to organize folklore songs to disseminate information about social desire. There are many examples in Bangladesh like the movement against British rule, the spread of the concept of fundamental democracy during Pakistan regime, even the recent Bangladesh independence movement was influenced by folklore song. Thus, sensor components are social elements or resource status, which create a condition for policy initiation or policy implementation. The publications like *Limits to*

Growth and *Our Common Future* are the sensor component of policies at global level.

5.4 ISSUES RELATED TO ACTORS

Policy actors are entities who influence policy formulation, execution, and/or implementation either to achieve the objectives of policy or to influence the policy outcome toward their own interest. Policy actors could be an institution, individual, or social organization. Functionally, the policy actors can be classified as political actors, policy formulators, policy implementer, policy executives, lobbyists, players, and policy interest groups. The sustainability of policy depends on how the actors function and their functional integration in the policy operation. Theodoulou and Kofinis (2004) mentioned two principles related to policy evaluation known as Wilson's law. The Wilson's law dictates that outcomes of policy evaluation depend on the nature of actor involvement. Sustainability evaluation may also encompass such problems. The Wilson's principles are:

1. Wilson's first law is that all policy interventions in social problems produce the intended effect—if the research is carried out by those implementing the policy or by their friends.
2. Wilson's second law is that no policy intervention in social problems produces the intended effects—if the research is carried out by third parties, especially those cynical of the policy.

Wilson's law dictates that the outcome of policy evaluation often depends on who is evaluating the policy. For example, oppositions in state politics always find some problem with policies of parties in power. With respect to sustainability evaluation, the following variations in policy operation in relation to the actor need to be identified.

5.4.1 Influences of Macroactors

In almost every country there are some organizations or individuals behind the scene of all political arenas who pass judgment on policy issues known as macroactors. For example, in many developing countries, the policy implication is affected by army influence and in some countries by the donor agencies, which are secondary in nature but very important. If such macroactors are many, their objectives may vary, and the objectives may often contradict with the core objectives of sustainability. Thus, policy evaluation should also look into the issues and interest of macroactors. Here, it is relevant to note that macroactors are not necessarily a social element but they influence policy. Conversely, microactors can be social elements but they do not necessarily influence policy.

5.4.2 Positioning of Actors

The idea of positioning is that actors have no definitive roles but are constantly being positioned in discursive exchange. Positioning requires some actions and at the same time actors who are not totally free from each other; that is to say both actions and actors will have to account for all sorts of preexisting understandings. The term can best be expressed if it is considered a sort of compromise where the role of both actors and actions are recognized but not legalized. Story lines also have their implicit positioning or they may function as the stimulants to achieve the policy objectives. Positioning can be an important technique of policy implementation in developing countries with huge populations, particularly where the difference of opinion is high. In policy evaluation, it is important because in most cases positioning for sustainability is oriented around actors from different fields, e.g., economic actors and environmental actions. Thus, positioning is a kind of mutual functioning.

5.4.3 Way of Arguing

Each policy domain appears to have a set of specific ways of arguing a case. Usually, a policy discussion follows through some format or some specific institution before being finally accepted. Thus, the arguing of a policy element uses social structure or institutions for appropriation of policy decision. Policy evaluation thus, in most cases, does not see what argument was occurring but how it had occurred. Way of arguing in the case of resource sustainability should be very much structured and well participated because it needs coordination with many policies of other sectors and subsectors. It is important for policy evaluation, when a certain unfavorable outcome of policy needs to be seen, whether it were as a result of inadequate structural arguing or not.

5.5 BLACK BOXING

Callon and Latour (1981) have introduced this terminology to express a policy process where things are made fixed, natural or essential for steering away the latent opposing forces. Producing information through story lines is a way of black boxing. In the policy process, the technique of black boxing is different from the technique of motivation in the sense that motivation is used for the general mass who do not know or not aware of any process. However, black boxing is a sort of generation of information for convincing the people that are aware and know things, so that opposition is minimized and coalition can be made. The knowledge used in motivation usually consists of the usefulness of the resources, whereas in the case of black boxing, knowledge is generated and consists of information of the past and future. Knowledgeable people are clever and their orientation has to be fixed in such a way that they cannot understand the motive. Thus, black boxing is a policy technique for special groups of social elements.

The discussion in the above paragraphs shows that policy elements influence social functioning or they themselves are influenced by social elements. Thus, the elements are discursive in nature and show why discourse of environment often combines with sustainability philosophies with a solution that does not match a sustainable format unless it runs into a political difficulty. Not thinking these formats through leads to unduly optimistic and technocratic thinking about policy change (not social).

CHAPTER 6

Components of Sustainability Assessment

6.1 INTRODUCTION

Perhaps, the greatest benefit of sustainability assessment is found, not in the direct results it generates, rather in the process of policy learning that accompanies it. In well-designed assessment processes, policy

Sustainability Assessment. DOI: http://dx.doi.org/10.1016/B978-0-12-407196-4.00006-4

actors learn constantly from the formal and informal evaluation of policies they are engaged in and are led to modify their positions on the basis of the information they collect and the knowledge they generate in this process. The lessons managers draw from evaluations, based on both objective facts and subjective interpretations of the facts, leading to conclusions about both the means and the objectives of different components of policy and are an essential component for the improvement of policy sustainability. The evaluation of resource policy has to consider certain social as well as economic criteria for sustainable options (GOB, 1995). Accommodating such criteria in policy evaluation requires comprehensive assessment of the sustainability climates of preexisting policies (OECD, 1984), which need to integrate information from trade, market, communication, biodiversity, and participation at local, national, regional, and global levels. As a result, sustainability assessment may appear as different components of policy evaluation. The following sections explain important components of policy evaluation related to sustainability.

6.2 SOCIAL ADEQUACY

Social adequacy of policy is considered as the judgment about policy competence or ability to meet the requirements of social aspiration and task-oriented demands of society for day-to-day business. Social adequacy of policy is more than social acceptability. A policy acceptable to society may not be adequate for sustainability. Some components of policies, like very fundamental traditions, may not be suitable for running under the given circumstances of other global and environmental situations of the time. Social adequacy involves human judgment as well as social capacity. Therefore, the uncertainty in policy sustainability due to social inadequacy could come from human beings as social elements and the capability of society to accommodate policy requirements.

In this regard, it can be mentioned that a sustainable policy designed for a developed country may not be socially adequate for a developing country, because the developing country may have a shortage of technological or trained personnel to meet the requirements of the policy. Social adequacy factors of policy and their implementation are generally considered as the background force for sustainability of a specific resource use type evolved from and within the society. Social inadequacy may result from an imbalance in either policy formulation or its implementation as a socially acceptable process. Perhaps, some of the most efficient policies in the developing countries are defective in their implementation due to lack of social adequacy (e.g., policies contradicting with social tradition).

6.3 SCIENTIFIC ADEQUACY

Scientific adequacy in relation to policy sustainability means that the arrangement or instruments incorporated in a policy are able to predict accurately and meet the future requirement of policy. Because policies are made for future actions, scientific adequacy of a policy denotes the progressive dynamism and flexibility incorporated within the policy to accommodate evolving processes of science and society. They are important components for policy sustainability. When policy factors are socially adequate for sustainability reasons, the information of those policy factors should also be scientifically adequate. For example, rates of land use change are often directly related to rates of population growth. However, increase in economic development generally diminishes land use locally (Houghton, 1994).

This signifies that knowledge from one type of science alone is not enough to judge the trends of a policy outcome. It requires motivation from political, social, and economic sciences consistent with long-term sustainability. The innovation of gene technology for improving growth and yield may be an excellent opportunity for supporting sustainability and is scientifically adequate. However, if the society itself is not in receptive mood, i.e., if the education and technological standard are not up to the standard for creating a provision for technological innovation, the policy measures may not bring a fruitful result. Conversely, we may say an event or issue which is scientifically adequate to create sustainability will have no meaning in policy sustainability if it is not acceptable to society. In fact, the actors will resist those events, though scientific, being included in social policy. For example, adopting the Kyoto protocol of climate policy is scientifically adequate to create environmental sustainability, however, for market economies of some developed countries like the USA the protocol is not desirable to society, at least a powerful component of the society, therefore, it has not been included in social policy.

6.4 STATUS QUO

Status quo determines the existing situation of society or nation. Status quo of a policy often means noninnovative and nondynamic situation, therefore, could have a negative impression on sustainability. However, status quo situations are often used for conservation purposes, which is usually considered mandatory for sustainability. In practice, the status quo situation is a third dimension that prevents damaging forces to work on social/resource elements for maintaining an existing adequacy. Along with social and scientific adequacy, sustainability consideration of factors of resource policy thereby depends on how the status quo situation of socioeconomic and sociopolitical changes are addressed.

Addressing the *status quo* situation of a policy involves many social issues such as sustenance, security, progress, equity, polity, and polarization. These issues have local, national, regional, and even global dimensions. But in a modern state, different issues are addressed by different policies. For example, in Bangladesh, about 56 acts have clauses related to management of the environment (Ali, 1997). Thus addressing/evaluating a single policy will not be adequate for maintaining the status quo in environmental policy. Similarly, trade or military policies might have linkages with the resource policy all of which need to be addressed for managing a status quo situation in sustainability. Thus, irrespective of social and scientific adequacy, evaluation of status quo addresses many issues *and* signifies the total set up of a policy within the society.

6.5 POLICY PROCESS

Policy process can be considered as social elements involving several stages like formulation, implementation, and modification or monitoring. In each of the stages, there are actors having different roles such as policy formulator, implementer, and players. They remain engaged with policy operations and modifications that determine policy adequacy and acceptability. When we say policy processes, we mean the continuous influences of actor activities at different levels related to policy metamorphosis. Thereby, it is plausible to assume that policy processes are prerogative activities of actors and hence policy sustainability largely depends on the coherence and sincerity of actor activities.

Almost all the policies of a nation are usually targeted to social welfare and macroeconomic requirements, but their outcomes vary depending on the role of the actors. Resource and environmental aspects in a policy involve a leading group of people for planning and implementing the idea. However, there could be players and political influential groups intervening between the planning and the implementation of policies for their benefit. As a result, a policy process though having adequate targets, may not have sustainability fate with regard to social outcome. Thus, process evaluation in relation to actor influence is meaningful for a sustainable policy. In case of policies already formulated and implemented, process evaluation can only be done if enough documents about the roles of the actors are available.

6.6 POLICY STIMULUS

Apart from actors directly involved in policy process, other actors associated with macroeconomic planning may become involved in the "policy coalition" of resource use. The activities of those coalition actors have been termed here as "policy stimulus." Policy stimulus is different from macroactor because policy stimulus actors work through coalition. Sometimes issues or achievements in other countries may become stimulus for sustainability. For example, success in a neighboring country in controlling population can be a stimulus for others. Such stimulus may be local, regional, or global. Strand and Toman (2010) have given an account of green stimulus for sustainable development. Though the roles of policy makers can be detected by process evaluation, the influential stimulus remains undetected. Identification of such stimulus is also important for sustainability evaluation of policy. Stimulus evaluation can also be considered as an ongoing process, evaluation thus has limited application in the past policy evaluation.

6.7 PARTICIPATION

A subservient policy usually addresses community issues to solve the problem. Policy evaluation, therefore, implies an attempt to develop a totally rational approach to the issues of community development. Although there are different types of participation, participation evaluation aims to determine the interest of community people toward the policy. Forest land use policy can be treated as one of the approaches of resource use attributed to community services, having relationships with many other factors of community development. Some of these factors have been described in the impact model (FAO, 1993). Community relationships with forest land use are mainly for fuel, food, fodder, fiber, shelter, income, and environmental services. Community participation designates how they are benefiting from the policy. However, participation evaluation is also an ongoing policy evaluation thus it has limited application in the evaluation of past policies.

6.8 SECTORAL GROWTH

Economic growth can be considered as one of the desired outcomes of policy, but it is hard to entangle a single policy with economic growth. Thus, sectoral growth would be the more relevant indicator of policy sustainability. Sectoral growth identifies the economic out turn from the sector; thus, its contribution to economic growth can be assessed. Atkinson et al. (1997) discussed the ways of measuring sustainability in detail. However, the benefit of looking at sector growth is that it would be easier to determine the sustainability by identifying whether the contribution to economic growth is short or long term. However, a policy could have contribution to different sectors; relating the resource contribution to all those differential sector growths would clarify the sustainability circumstances better than that of considering only the dominant sector.

6.9 RESOURCE EXPLOITATION

Evaluation of sector growth is not complete if the aspect of resource exploitation is not looked at. Exploitation itself means the scale of utilization in a sort of unreasonable way. For example, using forest resources for exporting unprocessed wood is a kind of exploitation— because the potential contribution of such resource use to social sector growth is minimal. Thus, the growth prospects of many developing economies are dependent to a considerable degree upon the performance of their resource exploitation mainly dominated on land-based production systems like agriculture and forestry (FAO, 1988). The impact of growth may be reflected on resource exploitation which may create an unsustainable situation. For example, the economic growth of Japan from the 1960s has reflected in the forestry sector by a sudden tenfold increase in paper consumption (Dargavel, 1992). This may be an unsustainable reflection; however, there are policy options to meet the situation.

Exploitation in the forestry sector can be explained in terms of exports, new road construction, urbanization, and a number of forest-based industries. On the basis of those reflections, a correlation might be drawn with the rise in income and increase in the consumption of forest produce. Correlation may also be possible by converting the consumption of forest produce into an equivalent land area required for supporting the increase. However, there could be more than one cause and effect that influence the forest resource use (Mather, 1990) which are different for different nations. Dargavel (1992) concluded that many of the causes are related to local and national policies. The multinational companies engaged in forest harvesting, mostly based in developed countries, encourage a certain form of growth and development and thus influence the forestry exploitation and sustainability. Thus, evaluation of policy relevance within resource exploitation is important for sustainability assessment.

6.10 TRADITIONAL PRACTICES

Traditional practices have often been blamed as the destructive process for resource use—particularly relating to inefficient use. Stebbing (1921) blamed traditional practices of India as a destructive process to forest land use. But they can be used as the sensor equipment of policy—particularly relating to conservative use. In many cases, traditional practices have been held responsible as destructive—not from the practice itself but from the intensity and/or practice of such practices. Although environmental geographers have done much to describe the relationship of environmental degradation and traditional practice, the roots of degradation are in fundamental dimensions of evolution (Boyden, 1987), population (Grainger, 1993), education, and consumption, the interpretation of which may be different between the political economists and the land use geographers. Traditional peoples have used forest lands of many countries for thousands of years without exhaustion of biodiversity and environmental characteristics. Policy consideration of traditional practices often determines the policy acceptability in developing society. Thus, evaluation of resource policies cannot go without including the status of traditional practice.

6.11 ROLE OF ACTORS

The different actors involved in resource policies, for example, related to forest land use and forest transitions are: multinational companies, national manufacturers, logging companies, workers and local residents, inhabitants and owners, international and local environmentalists. They involve the interplay of local, national, and international interests in the governments of countries, the mode of which usually depends on differentials of economic capacities and political power. All those processes raise political questions: who benefits and who loses in the process of forest degradation and how do the political systems of different countries deal with it? The purposes of sustainability assessment thus are to assess the comparative perspectives of the role of different actors as they relate to the resource policies. The actor role may be anticipated by examining production and trade in resources or resource-based products, industrial structure, processes of resource change, conflict with other resources, socioeconomic information, and finally by relating the achievements of national resource policies within the regional and international conventions.

6.12 FRAMEWORK ASSESSMENT

In a sound policy, usually several frameworks of actions are included, due the recognition to which should be given during the sustainability assessment. Most of those frameworks operate together to make a policy successful. Flaws at any stage of frameworks may be good enough to causing failure of policy sustainability. Table 6.1 shows some examples of policy frameworks.

For assessment of policy sustainability though the implication of all the frameworks may not be relevant, effort should be taken to evaluate as many frameworks as possible because investigation of resource policy and the environmental sustainability can be distributed across many frameworks of Table 6.1. Those frameworks will enable the location of the strengths and weaknesses of legal, judicial, and administrative criteria of policy.

6.13 SCOPE EVALUATION

According to Keys et al. (1991), the scope of resource policy sustainability usually covers contexts of many issues related to population

Table 6.1 List of Policy Frameworks	
Framework Type	Examples
Political framework	(i) Priority (ii) Government commitment
Legal framework	(i) legal directives (ii) Judicial
Legislative and regulatory	(i) Laws and acts (ii) Monitoring
International dimensions	(i) International agency network (ii) International conventions
Standards and institutions	(i) National (ii) International

growth, economic change, social change, political and legal perspectives, administrative reform, constitutional reform, urban planning, evolution of policy planning, contemporary planning systems, hierarchy and functions of plans, national and regional coordination, public participation, public rights to dispute, NGO involvement, and fiscal policy. Evaluation of some of these factors can be considered for sustainability assessment with other policy criteria as well. A few of the factors such as citizen's rights and urban planning may not be specifically meant for resource policy but have strong linkages with resource bases such as policies on land use. However, there are factors within the suggested list, such as contexts of administration, evolution of policy planning and hierarchy, which are important for sustainability assessment of policies.

6.14 EVALUATION OF IMPLEMENTATION

In addition to the context of a policy frameworks and scope, Mather (1986) has pointed out that certain other criteria should also be included for a sound resource policy; those are:

1. priority setting—e.g., economic (affected by land ownership pattern), social (employment or environmental),
2. education,
3. cultural factors, and
4. per capita income or expenses.

These criteria are notable for planning in agricultural and forest resource sectors. Particularly, the issues of priority setting are very important in the cases of developing countries because often the priorities set in writing are not reflected in implementation. The other factors like education, culture, and income are ancillary to priority setting. These are important criteria for successful policy implementation as well. Thereby, the total sustainability of policy is much dependent on those secondary social elements.

6.15 INSTRUMENT EVALUATION

Application of appropriate policy instruments largely determines the policy success. That is why environmental economists, Pearce (1990) and Repetto (1988), suggested policy intervention and financial compensation to prevent the rain forest destruction. Muzondo et al. (1990) and Nunnenkamp (1992) discussed alternative policies and a financial compensation scheme for that purpose. Binswanger (1989) and Mahar (1989b) have quantified some of the direct instrument effects (such as land tax and subsidy) on the rain forests, particularly rain forests of the Amazon, in a particular framework. Application of some of these instruments was in vogue in the resource policies of forest land use. Thus, it is important to consider the suitability of instruments used for policy implementation.

6.16 STRUCTURAL EVALUATION

The effectiveness of application of policy instruments largely depends on the structure of administration for policy implementation. Khuraibet (1990) suggested that decision-making in policy formulation could be evaluated by answering the questions:

1. Who has the power in the society to make a decision?
2. How are decisions taken?

Usually decision-making models explain how government makes decisions with regard to planning environmental or economic concerns. The theme of the models is the selection of one from many alternatives. Within a government such decisions can be taken in two ways:

1. rational actor model, and
2. organizational process model.

In the first one, government acts as a unitary body and investigates the alternatives of policy suggestions and selects the most rational alternative. In the organizational process model, government asks all its departments or ministries about a particular problem and ultimately a decision is taken with the support of the majority of organizations. But in the political bargaining, a different approach can be taken in decision-making. Instead of specialized bodies, in political bargaining the decisions are taken by players within the government who have the ability and credibility to persuade government to take certain decisions on the subjects under debate. Thus, the nature of policy depends on the structural components of a policy.

6.17 CAUSE EVALUATION

The causes of resource use criteria and environmental degradation depend on factors like ownership and policy purpose. The criteria, factors, instruments, and mode of policy implementation vary from nation to nation and from time to time. Conversely, the outcome of the same policy may be different in a different space at a different time. Using the Computable General Equilibrium (CGE) model, Thiele and Wiebelt (1993a) extended a forestry submodel along the line of Dee (1991), which may allow examination of conventional forest policy instruments, e.g., resource taxes, to secure criteria like property rights, selective logging regimes, and the setting up of national parks. Criteria that the model captures from the implication of environmental and economic policy instruments for resource use patterns seem to be important because the question of whether the resources should be harvested or left as protected or cleared entirely for production of other resources is primarily a question of resource use pattern. If the common guideline for sustainability can be incorporated within the policy, then a selection of alternatives can be targeted on the basis of economic criteria only.

6.18 COST EVALUATION

Policy decisions can be evaluated by IRR or cost–benefit efficiency (Khuraibet, 1990). But they require some economic valuation of the decision. Under the United Nations approach, it is considered that issues of basic human rights, such as drinking water, primary health, and basic education, do not require economic justification. The situation of environmental sustainability is also not always measurable in economic terms. An environmentally sound policy promotes indigenous capacity and peoples' participation in the policy process and awakens consciousness. There is no standard way for measuring these variables. Beneficiary assessment on the decision may be one technique of policy evaluation but such evaluation is often targeted to lower income group, whereas there is a possibility that the lions' share of the benefits of resource policy accrue to the dominant groups. However, if costs are considered as a calibrator of the scale of forestry operation, environmental status may be understandable by comparing the cost of investments at different time. If the costs are known, the cost calibration can also be used for the evaluation of past policies.

6.19 IMPACT ASSESSMENT

Environmental impact assessment may be another process of policy evaluation for the environment. The impact evaluation may be possible by counting the opinions of the affected people or participants, thus the impact assessment of the policy has been termed as social evaluation of policy (Valadez and Bamberger, 1997) and is widely used by the Asian Development Bank (ADB) for program analysis. Such evaluation is convenient to conduct before the implementation of the policy looking at how different groups within the society will be affected by a single or a few decision of policy; however, a policy may have many decisions and policy level impacts may affect a whole nation. Under the circumstances, it will be difficult to assess the policy impact based on the opinion survey at national level. Further, application of this process is also not possible for assessing the policies of the distant past after which many incidents must have happened or the participants are absent from the society.

6.20 QUANTITATIVE APPROACH

Considerable progress has been made on the qualitative approach of policy evaluation but long-standing disputes between the advocates of qualitative approaches and quantitative approaches have not been resolved. However, quantification as a rapid assessment processes is now getting priority for its cost-effectiveness and it is most responsive to the sociocultural environment (Kumar, 1993). For example, rapid rural appraisal has drawn the interest of many project proponents. The holistic approach is another process, which enables to understand the interaction between the policy and the social cultural, political, and economic environment (Marsden and Peter, 1990). Although some progress has been made on these processes at smaller scale evaluation, the evaluation of policy perspective has not yet been integrated fully into the monitoring and evaluation system. While there is a belief that NGOs can contribute to a greater extent to the holistic approaches and qualitative analysis of policy (Marsden and Peter, 1990), the approaches so far have made little impact on the governments of developing countries (Valadez and Bamberger, 1997). For these various reasons, the methodological system for policy evaluation has not been systematized and developed as a conventional analysis method.

6.21 ANTHROPOGENIC EVALUATION

Anthropogenic influence is considered as one of the real problems of sustainable management; therefore, systematic assessment of it is important for policy evaluation. Anthropogenic influence on the forest resource system can be based on the population (determines the size of consumption/influence), affluence/income (determines the level of consumption/influence), and technology (determines the rate of consumption/influence). On the basis of these influences, efforts have been made by population biologists, ecologists, and environmental scientists (Holdren and Ehrlich, 1974) to establish the relationship between human welfare (income) and environmental impact and put forward the proposition of the well-known IPAT model. The model postulates that environmental impact (I) is the product of population (P), per capita affluence (A), and technology (T). This model discusses the principal factors of anthropocentric influences, known as "driving forces" of environmental change (Ehrlich and Ehrlich, 1990; Ehrlich and Holdren, 1971; Holdren and Ehrlich, 1974). The model assumes that:

1. Population is usually considered as the key driving force along with its economic activity, technology, political and economic institutions, and attitudes and beliefs (Dietz and Rosa, 1994; Mather and Needle, 2000).
2. A number of adjustments to the population are possible, such as level of education, gender, skill, and those treatments can be used as the orientation perspective of the evaluation discussion.

3. Little effort has been made to discipline the model since its inception two decades ago (Dietz and Rosa, 1994).

In particular, social scientists and geographers generally have ignored the model while biologists, ecologists, and other environmental scientists generally considered that the proposition is almost true, therefore, have not been motivated to test it rigorously. It can also be argued that the IPAT model can be taken as a plausible means for bridging the difference between social and biological sciences on the historical and contemporary problem of environmental sustainability. Thus, the ultimate aim is to generate more disciplined study but less debate on policy evaluation which is not necessarily grounded on empirical research.

Much of the debate about population, affluence, and technology on the environment can be structured by the IPAT model and it can be widely adopted in environmental evaluation of policy but the model has also some plausible limitations. The main among these is that it does not provide an adequate framework for disengaging the various driving forces of anthropogenic environmental changes. As a consequence, the IPAT model guides the effort of evaluation toward a cumulative theory of empirical findings (findings are combined effect of all the factors). If it is possible to sketch alternative ways of conceptualizing the driving forces of the anthropogenic changes, and to look at some additional forces other than population, income, and technology, it may be possible to propose a reliable change in the model rendering it more amenable to empirical separation (so that a single factor can be emphasized in the prediction of environmental sustainability). In practice, evaluation like participant evaluation, sector growth evaluation, and impact evaluation may be the component of the IPAT model.

6.22 INFLUENCE OF OTHER POLICIES

From the views discussed in the above paragraphs and according to much of the literature (Thiele and Wiebelt, 1993b), the causes of resource degradation can be correlated with a bunch of policies rather than a single policy. Thiele and Wiebelt (1993b) considered the theoretical framework for quantitative estimation of likely consequences of policies to reduce resource degradation. The evaluation of resource use thus needs to be approached in a holistic view. Two approaches can be suggested:

1. Bringing the policies, which are causal of resource degradation under a comprehensive macroeconomic model, maximizing the economic benefit, in favor of which Bôjo et al. (1990), Bolton (1989), and Devaranjan (1990) discussed in their works.
2. Looking at the bunch of policies under a modular environmental approach, maximizing the environmental benefit.

In practice, the main target of the policies is either the national fiscal management or the environmental management. But an optimum incorporation of both the issues within a policy is important for sustainability. However, this is a special case to look at influence of a bunch of policies, thus accommodation of such optimization is not important for the purpose of this book. Nevertheless, the proposition of assessing the influence of other policies leads us to evaluate the linkage of policies.

CHAPTER 7

Linkages of Sustainability Assessment

7.1 INTRODUCTION

Linkage evaluation identifies the relationship between different policies or different decisions of the same policy. For example, decisions in government functional system can be tripartite: executive, legislative, and judicial. All the three functional systems are involved with the cumulative and distributive issues that are associated to policies, therefore, have strong roles for sustainability. Evaluation of such linkages is important to investigate the dominance of one policy component over the other. Often a policy decision taken from outside the policy area may be detrimental for the resource and environmental sustenance.

Sustainability Assessment. DOI: http://dx.doi.org/10.1016/B978-0-12-407196-4.00007-6

For example, decision on land use in agricultural policy may be detrimental to forest land use. Daniere and Takahashi (1997) indicated that where urbanization or urban migration is high, environmental policy might be hampered due to government attention to other policies linked to maintain the problem of migration and arrangement of urban amenities. Fiscal policies, exchange rates, terms of concession of public land, price controls, transport networks, land and tree tenure, tariff and nontariff barriers to international trade, investment incentives, agricultural sector's strategies, and other macroeconomic policies may affect economic motivation in the management and conservation of forests. Therefore, the influences of the cross-sectoral linkages, linkages among cultural values, attitudes, and behavior are important to the design and implementation of sustainability aspects of a policy. However, consideration of the strength and nature of linkages are also important for deciding the justifiable level of investment and admirable effort of sustainability assessment. Bamberger (1991) has described sectoral policy areas, specific types of instruments influencing the policies, linkages affecting forestry, possible effects on forest development, and option for actions. Depending on the aspect and strength, the following linkages of forest land use practice with other policies/sectors may be considered for evaluation.

7.2 PARALLEL LINKAGE

Parallel linkages show how resource sectors are directly related to other sectors, e.g., environment and forest land use, tourism and forest land use. Such linkages are also known as sectoral linkage. Parallel linkages generally involve sectors related to a single frame— here it is land use. For example, tourism and forest production may be developed on the same frame of land. Assessment of such linkages gives equivalent emphasis of the sectors and may include aspects of common usage, economic output, and interdependent relationships.

7.3 LINKAGE OF ASCENDANCY

This linkage involves the evaluation of policy issues, which are not related to the same frame. This sort of linkage is more common in policies that influence the frame. This means, instead of considering a common frame, one of the policies considered in linkage analysis could be a frame policy itself. For example, land use and tenure. If the linkage of forest resource is considered with land tenure policy (not land use policy), then the relationship will be a linkage of ascendancy. This is also called tenurial linkage. In those cases, decisions may need to be made on a priority basis. The common denominator of almost all resource-base policies is the land resource; as a result, land tenurial aspects overlap. This means the tenure may be considered as a frame policy. For example, oil exploration, coal mining, or agriculture within the forest area produce land tenurial conflict. Those conflicts need to be addressed in the analysis of land use whether the conditions for use of land for one aspect could ensure the conservation or reestablishment of other aspects as well.

7.4 LINKAGE OF DESCENDANCY

Linkage of descendancy mainly developed when requirements of one policy imposed and influenced the decision of other policies unrelated to the frame. For example, in forestry cases, they might happen from the competitiveness of the investment in the forestry sector resulting from the variation of economic efficiencies set out in the policies or changing in the value of currencies. Thus, the linkage may be named as economic linkage as well. As an example, deMontalembert (1995) indicated that the actual rent received by the Filipino government during 1979–1982 from the forest was much lower than the potential rent. As a result, to accrue the targeted revenue, more forest than the expected area was felled. Thus, economic incompetence of policy implementation descended down to extent (area) of forest land, use affecting sustainability.

7.5 LINKAGE OF HIERARCHY

Linkage of hierarchy connotes the linkages between the administrative bodies of organizations under the same policies or different policies. This may also be termed as administrative linkage. For example, in Bangladesh, a forest intruder, if caught, is handed over to the police and thereby an efficient control of illegal activities within the forest requires help and cooperation from the police department. This is not a policy linkage between two policies as such but a linkage between two administrative bodies that ensures the success and sustainability of forest policy.

7.6 HORIZONTAL LINKAGE

Horizontal linkages between institutions strengthen policy measures. They are different from the hierarchic in the sense that the requirements of such linkages do not have immediate effect and there may be an alternative (e.g., alternative institutes) to choose for establishing linkages. They are also not parallel, because they are not linkages for the same frame used by a different sector, but linkage of a different institution may be from a different or same sector. It may also be termed as institutional linkage. Sometimes, weaker attention to such linkage may cause a setback in the implementation of policies. For example, an institution preparing professionals for forestry management and administration needs to be emphasized equally within the Forest Policy arena, otherwise a fatal flow in implementation may develop due to lack of manpower hampering the total achievement of a good policy. Regional organization and treaties like EU, SAARC, ASEAN within the global arena and the individual countries within the regional treaties are examples of horizontal linkages. According to De Bruijn and Norberg-Bhom (2004), policy sustainability often depends on strength of bonds among the countries of such regional organizations.

7.7 QUASI-POLITICAL LINKAGES

Linkages of policy with affairs like transmigration, subsidization, aborigine issues, stakeholders, and international pressure is neither fully economical nor substantially political but a balance between the two. They cannot be considered under horizontal linkage because they are not regular policies but political. Formation and formulation of such policies are different than regular policies. Occurrence and recurrence of green politics in developed countries demonstrate coevolution of policies with sustainability. Even the liberal and democratic policies though have different commitments, nowadays they essentially keep an environmental component in their policies to demonstrate sustainability concern. Quasi-political linkages often provide disguised and multidimensional roles in affecting policy sustainability.

7.8 EXTERNAL LINKAGE

In developing countries, resource policy is predominantly targeted to economic development; therefore, it is expected that resource policy will be highly related to other policies relevant to the national economy such as: fiscal matter, trade, infrastructure, industry, and investment (also criteria evaluation). Many such policies are targeted mainly to alleviate the economy and to reduce the pressure of recession and/or debt crisis. In addition, a new wave of global environmental concern, particularly relating to climate change and environmental protocols, drives sustainability forces in resource and environmental policies of countries. However, there had been withdrawal of countries from the global commitments, denial of ratification. The commitment of countries linked with such external forces determines the mix of policy sustainability.

7.9 MARKET LINKAGE

Most evaluation of public policies concentrates on economic analysis and occasionally to consequences of market failure. In the resource sector (e.g., forestry), market failure is incomplete but extensive and could be a common issue. Usually, government intervention faces market failure. As government's intention is often motivated by politics, policies are developed from such a type of intervention. Rausser (1992) has referred them as "Political Economic Resource Transfers" (PERTs). In designing and implementing PERTs policies, markets are viewed as separate from the political process. Thereby, such linkages seem to be external, their influence work through market mechanism.

7.10 EVALUATION OF LINK TO THE PAST

Narratives about the past history of a policy produce a story line, which allows actors to draw upon various discursive categories to give meaning to specific physical or social phenomena. Drawing the past into the assessment plays a role in the positioning of subjects and structures. The key use of historical comprehension is to understand the different components of the problem at its time boundary. Policy changes may take place through the emergence of new story lines that reorder the understandings. Thus, finding an appropriate story line becomes an important form of agency for the sustainability assessment of policies.

7.11 ACTORS AND STORY LINE

The construction of story line (knowledge) and actors involved in a policy are important for the evaluation of linkages of past activities or policies. Actors are often involved in transforming academic information into policy practices. Within the policy processes, there are policy brokers who translate the knowledge to policy information. For example, taking an early measure to abate the environmental degradation was not possible in the Netherlands because it failed to translate the scientific information of the study of acid deposition into a policy. However, the scientific information was brought to a "problem closure" by social repercussions that developed out of it (Hajer, 1995). This signifies that sustainability in a society cannot be assessed only on the basis of the present social system. On the contrary, society can be interpreted as a system of transformation drawn from the story line of the past and dilemmas of practices of actors in different arenas.

7.12 PRACTICES AND STORY LINE

Story lines of past policies can also be constructed by analyzing the activities, utterances, and practices of actors, but the relative independence of a story line from the actors opens up new ways of understanding the problem. Story lines maintain the structure of the controversy, but individual freedom to change discursive contributions in actual practices allows the analyst to safeguard the credibility of expression. For example, credibility may be changed with the changes of the chair (e.g., the CCF for forestry), or by the introduction of a new definition (e.g., scientific forestry). For this reason, story line and specific discursive interaction are better evaluated as tied to specific practices rather than actors. Thus, sustainability assessment of policy is not what is being said but essentially what is being practiced.

7.13 REFLECTION OF IMAGE OF CHANGE

In the resource sector, a particular role of policy can be fulfilled by the creation of an image of change. The image of change can be outlined by characterizing the status of conditions shown in Fig. 7.1. The figure shows the way of assessing the changes of environment in policy evaluation. It is an integration of environmental assessment and socioeconomic assessment.

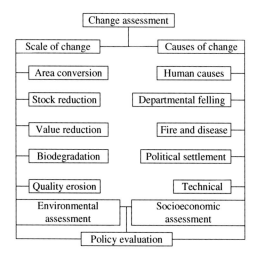

Fig. 7.1 Policy evaluation through outlining image of change.

However, the discursive features in Fig. 7.1 may differ in different countries. Hence, although it is fair to say sustainability is an international policy discourse, one really has to allow for national particularities through the reflection of environmental and socioeconomic images as shown in Fig. 7.1. The unique particulars of discursive elements may be selectively generalized if the issues of a policy are processed and pursued through certain phenomena. The following sections attempt to present some steps of discursive phenomena.

7.14 INTEGRATING INFORMATION

Integration of information is one of the important factors for decision. If the information is scattered, rational decision is not possible where consideration of all the factors can be given. According to Osleeb and Kahn (1998), a policy decision deals with a large size of national and historical data which may be integrated by:

• Geographic Information System (GIS),
• Spatial Decision Support Systems (SDSS), and
• Spatial Models.

Organizing information helps in establishing linkages and selecting priorities in indecision. The use of modeling tools may help in generalizing discursive issues of policies; however, the precision in sustainability achievement largely depends on how the decisions are made.

7.15 FORECASTING

Forecasting is a technique for laying down options. At the beginning, the options are identified and the list of options can be narrowed down initially by forecasting the future and then detailed assessment of options can be made. According to Armostrong (1998), forecasting may be based on:

- judgment,
- extrapolation,
- econometric models, and
- combined forecasts.

The most important aspect of forecasting is to communicate the assumptions made and degree of certainty behind the forecast.

7.16 ASSESSING OPTIONS

For achieving a certain goal, a policy may prescribe particular option from many on the basis of assessing the options for their risk and cost. Merkhofer (1998) says that the common tools for assessing options are:

- probabilistic risk assessment method,
- cost−benefit analysis, and
- decision analysis.

Each analysis performs a specific function; therefore, more than one form of analysis may be needed for an assessment of option.

7.17 POST-DECISION ASSESSMENT

"Looking back and evaluating the outcomes of a decision, one may have a better basis for choosing the future action" is a wise saying in post-decision assessment of policy. According to Bergquist and Bergquist (1998), the common tools of post-decision assessment are:

- indicators and indicator system,
- goals and goal systems,
- budget-accountability system, and
- program evaluation system.

Post-decision assessment can be conducted focusing on whole or different components of a policy as well as different parts of an organization. The efficiency of different types of tools used for post-decision assessment depends on the components, purpose, and stage of assessment. For example, indicator systems are flexible tools that provide concrete, precise evidences, and focus on results. However, budget-accountability systems allow reallocating resources to areas of need or higher priority. It uses the socioeconomic instruments.

Leeuw (1991) thought that the dimension of policy discourse is to consider where things are said, and in what specific ways the things can be structured or embedded in the society. Thus, discourse analysis can be used as a specific ensemble of ideas, concepts, and categorizations that are produced, reproduced, and transformed in a particular set of practices and through which a meaningful interpretation could be given to physical and social realities. There could be numerous practices that can exert influence on sustainability issues, but the assessment of policy issues with sustainability is mainly relevant to the image of damage, role of science, and the role of regulation. Thus, addressing sustainability issues will be more effective, perhaps, when both problem environment and economy are addressed together. The whole sum action of addressing sustainability in a policy emblem could be as follows:

- Internalization of problem; aiming at bringing changes within the target group.
- Determining priority from which problem-solving measures should be started.
- Defining the parameter within which a solution was to be found so that regulation of environmental management becomes more effective and at the same time revitalization of economy is affected.

Hajer (1995) explained that the internalization of problems and priority setting of the discursive order (like laws, organizational routines, or categorizations) requires a constant and wide reproduction to guarantee the continuity of its meaning structures. This implies that the institutions and argumentative actions are to be examined in their interrelationship. That means practices cannot be evaluated as fixed entities, because their meaning is likely to change over time. The coherent argument also implies that policy change can be materialized only if one succeeds in finding ways to overturn routinely reproduced cognitive commitments.

CHAPTER 8

Assessment of Policy Instruments

8.1 INTRODUCTION
8.2 APPROACHES OF IMPLEMENTATION
8.3 ATTRIBUTES OF INSTRUMENT
8.4 CHOICE OF INSTRUMENTS
8.5 INSTRUMENTS AS A COMPONENT OF POLICY DESIGN
8.6 ADDRESSING THE IMPLEMENTATION OF INSTRUMENTS

8.1 INTRODUCTION

Policy instrument is a linkage between policy formulation and policy implementation. The intention in policy formulation is reflected in policy implementation through instrument. Policy instruments are often known as governing tools as well, particularly when they are applied with all conditions associated to them. The implementation of governing tools is usually made to achieve policy targets of resource management but adjusted to social, political, economic, and administrative concerns. Thereby, concerns of sustainability largely depend not only on what instruments are selected but also on how they have been applied. Assessment of policy instrument thereby can be an important component of policy sustainability.

8.2 APPROACHES OF IMPLEMENTATION

Implementation of policy instruments involves stages of selection, application, monitoring, and adjustment. Selection of instruments is made from best alternatives of a list of suitable policy instruments. They may be different for different component of policy. They are applied and monitored associated with governing regulations and adjusted with

Sustainability Assessment. DOI: http://dx.doi.org/10.1016/B978-0-12-407196-4.00008-8

political and social variables over time. There are different approaches to how instruments can be implemented outlined as follows:

1. The policy process is desegregated into a series of stages and appropriate instruments are identified with the different stages of policy process.
2. The large range of implementation techniques available in a government can be classified into a generic class.
3. Finding the actual instances of the instruments used. Explanation may be given for why a particular government chooses one kind of instrument over others.

According to Howlett (1991), generally two common approaches are used for dealing with policy instruments, the "resource" approach and the "continuum" approach. In the "resource" approach, various techniques are categorized according to the nature of the governing resources they employ; for example, fiscal resources or organizational resources. In the "continuum" approach, instruments are ranged against some choice which government usually make in the implementation process, for example, whether to use state- (i.e., regulatory) or market-based instruments. Both the approaches may have a separate starting point in the policy implementation process; hence, they are not mutually exclusive.

In the case of "resource-based instruments," either fiscal or organizational resources, an assumption may be made before evaluating a policy that "the instruments are not entirely substitutable, i.e., particular instruments have particular capabilities and particular requisites." Such an assumption will justify the evaluation of policy in regard to their instruments, whether the instruments chosen by the government have matched the job they are expecting from the performance of the policy. Another perspective of such an assumption in policy evaluation is to look into the established technical specification of each instrument made by the government to see their theoretical suitability for addressing a particular problem targeted by the policy. Considering the suitability of those technical specifications, the choice of a particular instrument can be justified on the basis of parameters of a given policy situation and matching the needs for a requisite action. At the same time, the evaluation may look into the supply of government resources required with the capabilities and the resource demands of the governing instruments.

Another assumption may be that there are similarities between instruments; and instruments are technically substitutable. This assumption may be considered as a component of the "continuum" approach. Under this approach, the rational choice of policy instrument becomes difficult and the choice mainly depends on factors outside the conventional policy process such as political choice. Under those circumstances, different governments may tend to choose different instruments from the same or similar policy. When both the approaches are considered together, it can be summarized that the "resources" approach considers technical aspects and the "continuum" approach considers the contextual aspect of policy instrument choice. Technical aspects of policy instruments are mainly socioeconomic, whereas those of contextual aspects are mainly sociopolitical. Therefore, evaluation of policy instrument can be expressed on the basis of social, political, and economic attributes.

8.3 ATTRIBUTES OF INSTRUMENT

Policy instruments may have individual attributes on which the choice of instruments may depend. Linder and Peters (1989) recognize that there are four basic attributes of instruments, which are usually considered by government in selecting a policy:

1. Resource intensiveness, including administrative cost and operational simplicity.
2. Targeting, including precision and selectivity.
3. Political risk, including the nature of support and opposition, public visibility, and chances of failure.
4. Construction on state activity, including difficulties with coercion and ideological principles limiting government activity.

These attributes are related to social, political, and economic characters of the society and, therefore, cannot be evaluated ignoring the contextual linkages with the social attributes. Considering the complexity of social attributes, policy analysts can arrive at some mix of ideas to determine whether selection of an instrument in the policy was social or political. It is worth noting here that assessment of social instruments may need to adopt a different approach than that of political instruments.

8.4 CHOICE OF INSTRUMENTS

The key determinant of instrument choice in developing countries could be dominated by the pursuance of economic objectives rather than that of environmental objectives. Decision-makers usually weigh the different options available and perform their own kind of administrative cost— benefit analysis in final choice of instrument. If that cannot be done, instrument choice is ultimately a political decision, heavily influenced by the nature of beliefs, attitudes, and perceptions held by political actors. For example, the present situation of forest policy instruments in Indonesia, in a structural context and conjectural context, is serving only short-term economic objectives determined by political parties, their representatives in the government, administrators, and interested groups. Such a choice puts limitation on institutional capabilities to emphasize technical capacities of specific instrument.

On the basis of a "continuum" approach, Howlett (1991) explained about some technical capacity of instrumental choice. Those factors are presented in Table 8.1. The table shows that the continuum approach of instrument implementation is not a single decision but a

Table 8.1 Factors of Instrumental Choice Under Continuum Approach of Policy		
Continuum	**Factor**	**Scale**
Continuum-1 Government to private	Nature of instrument ownership	Government—part government—joint government—private—private
Continuum-2 Compulsion to persuasion	Nature of government influence	Compulsion—arbitration—mediation—conciliation—information
Continuum-3 Direct to indirect	Nature of government control	Nationalization—licensing—taxes—subsidies—macromanipulation
Continuum-4 Compulsory to voluntary	Nature of instrument membership	State—institutional—organizational—company—private clubs
Continuum-5 None to full	Nature of instrument autonomy	Bureaucratic—semiautonomous—autonomous—independent
Source: Howlett (1991).		

set of decisions placed in a different arena. The outcome of policy depends on the whole set of decisions. Thus, corruption at any of the arenas may hamper the policy outcome. Also, there is a significance of choice of arena where the highest emphasis can be given for instrumental implications.

8.5 INSTRUMENTS AS A COMPONENT OF POLICY DESIGN

Policy instruments have been studied in the USA as a policy design approach that synthesized both the "resources" and the "continuum" models for selection of instruments. The study identified a certain basic number of general categories of instruments based on governing resources. The study also identified a number of continua describing the kinds of choices governments made in selecting instruments. Linder and Peters (1989) have cited a few policy instruments commonly used by the governments shown in Table 8.2.

The choosing of instruments may vary from government to government depending on the nature, authority, and priority. Their intensity may also vary depending on the scale of application as shown in the continuum approach (Table 8.2). Thus, though different governments choose the same instrument, the output of policy may vary depending on the continuum scale. Therefore, the implementation design of policy instruments depends on the points of technical choices shown in Table 8.3.

Table 8.3 shows that the design criteria of policy instruments are related to the criteria of policy climate described before. Implementing all the criteria on a good scale may not be possible for designing policy

Table 8.2 Common Instruments Applied by Governments for Policy Implementation	
Motives	Instruments
Financial	Loan, tax, fee, charge, fine, insurance, and price
Promotional	Cash grants, loan guarantee, public investment, government provision, public promotion, and kind transfer
Motivational	Information, demonstration, and government-sponsored enterprise
Regulatory	Quality control, guideline, prohibition, quota, and ban
Administrative	Certification, screening, license, permit, lease, and contract
Source: *Linder and Peters (1989).*	

instruments but the evaluation of policy instruments depends on the result of the combined effect of all the criteria. If some of the criteria are on a good scale, e.g., complexity of operation or reliance on market, the chances of failure would be automatically on a good scale.

However, Howlett (1991) reviewed Hood's (1986) work on policy evaluation where he shared the idea of technical substitutability. Hood's argument was that the design of policy instrument essentially depends on the four responsibilities/resources at the disposal of the government: Nodality/informational, Authority/coercive, Treasury/financial, and Organization/institutional (NATO). These resources can be categorized as detectors or effectors, originating the eight differentiated categories of instruments presented in Table 8.4.

The categorizations of government resources presented in Table 8.4 clarify the resources presented in Table 8.2 that some of the policy implementation measures may be for detection and some other for effective control. Monitoring function uses the information from detection measures, and thus helps the effective implementation of policy.

Table 8.3 Criteria of Policy Design and Evaluation and Their Benign Scale

Serial Number	Design Criteria of Instruments	Scale (good–bad)
1.	Complexity of operation	Low–High
2.	Level of public visibility	High–Low
3.	Adaptability across users	High–Low
4.	Level of intrusiveness	Low–High
5.	Relative costliness	Low–High
6.	Reliance on market	Low–High
7.	Chances of failure	Low–High
8.	Precision of targeting	High–Low

Source: Linder and Peters (1989).

Table 8.4 Categorization of Government Resources for Policy Implementation

Category	Nodality	Authority	Treasury	Organization
Effector	Advice	Laws	Grants/loan	Service delivery
Detector	Survey	Registration	Consultants	Statistical bureau

Source: Howlett (1991).

However, instrumental evaluation is made to see specific points where the instruments are malfunctioning. Covering all the aspects of instrumental implication in evaluation may be costly and time consuming. There could be a limitation of availability of information particularly in evaluating policies of the past. Thus, instrumental evaluation depends on the purpose of policy implementation it is addressed for.

8.6 ADDRESSING THE IMPLEMENTATION OF INSTRUMENTS

Dias and Begg (1994) argued that evaluation of a policy and its implementing instrument depends on the type of policy and its relationship with the environment. Also such evaluation of instruments depends on whether the instruments used on environmental problems occur in the natural resource sector, and are national, regional and global issues. Common approaches through which environmental problems are addressed or instruments are implemented can be grouped as shown in Table 8.5.

In a policy, some of these approaches are encouraged to have appropriate information about the environmental issues, and thus decision on an appropriate instrument becomes easier for the government. The efficiency and effectiveness of instruments in achieving the target of the policy largely depend on the efficiency of the approaches used for addressing the problem. Also, the instrumental approach and efficiency depend where they are targeted. In the case of forest policy, the instrumental targets could be:

- domestic consumer,
- domestic investment,
- marketing quota and control,

Table 8.5 Approaches for Addressing Instrumental Issues in a Policy		
Control Approach	**Enforcement Approach**	**Optional Approach**
(a) Command and control regulation 　(i) Setting standards 　(ii) Quality standards 　(iii) Process standards (b) Economic incentives and disincentives (c) Environmental impact assessment	(a) Use of discretion (b) Administrative mechanisms (c) Criminal penalties (d) Criminal litigation (e) Right to know, to inspect (f) Duty to disclose	(a) The role of general public (b) Environmental organizations (c) Corporate responsibility (d) Industry pressure groups (e) Environmental audit (f) Industry self-regulation (g) Narration of development and environment with conclusions

- revenue increase,
- foreign trade, and
- conservation.

The policy characteristics depend which of the targets has been emphasized. It is notable here that the targets of most past forest policies did not emphasize the environment (FAO, 1988). In those cases, an additional target of environmental management needs to be surrogated and then evaluations need to be done on how the environmental purposes have been served through other targets of the policy. Thus, the social dimension of the past forest policies could be evaluated to see how their execution ended up in the present physical and environmental condition of the forest.

CHAPTER 9

Social Perspectives of Sustainability

9.1 INTRODUCTION

Evaluation of social dimension of a policy involves the roles institutions or organizations play in policy process. The policy decisions are strongly influenced by the power groups especially in the developing societies. The influences often bring positive (coalition) or negative results (conflict) on the policy process. Three authors of three different decades, Mills (1956), Rose (1967), and Fairweather and Tornatzky (1977) have characterized those issues either as an extension of frontier psychology or a manifestation of capitalist economics. They may also be the legacy of autocratic ruling or colonial power. The worst thing of all for a sustainable society is the increased striving for power for its own sake and a displacement of original goals and purpose. Another characteristic of evaluating the social dimension is to understand the behavior of people. Social, political, and economic information helps to understand people and the ways that a policy decision is going to affect them (Freudenburg, 1998). The basic tools for characterizing the social, political, and economic settings, as recommended by the US

Sustainability Assessment. DOI: http://dx.doi.org/10.1016/B978-0-12-407196-4.00009-X

National Center for Environmental Decision-making Research (NCEDR), are:

- Secondary/archival techniques
- Primary/field work techniques
- Gaps-and-blinders techniques.

These tools are developed within the social sciences over decades and provide many avenues for information gathering. In the case of sustainability assessment, often the traditional tools of social sciences need to integrate with contingent valuation, valuation of resources or factors that are not actually sold or bought in the market. The skill and aptitude used in contingent valuation helps in making rational decisions on the environment. Thus, the sustainability components of a forest policy as such are not a part of compact policy but are integrated with individual policy programs of society. Therefore, evaluating social perceptions of a particular policy is to evaluate its status and linkage within the arena of the society. This share usually depends on:

- Who has formulated the policy?
- What lasting social benefit or cost could the policy produce?
- Who will derive the benefits?
- What environmental problem may be created?
- What alternatives do exist?

Usually, there are power groups as well as pressure groups for environment and sustainability who influence the policy situation within the society. Power groups, in general (have power, may or may not be involved), and pressure groups (involved, but may or may not have power), in particular, of developing countries are usually from higher income levels. They sometimes use the environment as a tool for competing for political power. This is also similar in state-level activities; rich nations put pressure on poor nations to maintain environment, whereas poor nations want something else, such as economic development. However, the difficulties in describing the problems are seemingly an unending struggle for achieving social and economic power by contemporary groups and institutions, often without regard for long-term effects on the social and physical environments. Thus, sustainability consideration of a policy embraces social events that need to comprehend certain aspects of a number of social variables which Fairweather and Tornatzky (1977) called "social situational variables."

Table 9.1 Social Situational Variables	
Internal Processes	**External Processes**
Organizational components	Social climate
Hierarchical structure, size, complexity, formality, and informality	Socioeconomic indicator
	Measurement objectiveness
Group dynamics	Geographical location
Cohesiveness, norms, leadership, composition, morale, and reinforcement	Folkways and mores
	Publicity and media exposures
Fiscal process	Relationship to other organisms
Income, costs, rate of pay, and book keeping	Legal constraints
Membership	Time
Voluntary, involuntary, and turnover	
Source: *Fairweather and Tornatzky (1977)*	

Social situational variable can be an "internal" or an "external process." Social situational variables are presented as a list in Table 9.1.

The internal situational variables shown in Table 9.1 are not independent of one another. There are common dimensions representing and associated with each variable. A variable, e.g., organization, may have different reinforcement systems and a different status of role relationship. It can define the differences between the internal processes of the policy and may have a relationship with some dimensions of enforcement under "group dynamics."

However, among those there are variables that are relatively easily manipulatable (controllable) than the others. Some of the variables are presented as examples and classified according to their relative manipulation characteristics in Table 9.2. When a policy measure is implemented, usually it should look into the manipulatable variables to achieve the nonmanipulatable variables. On the contrary, if a policy measure tries to achieve improvement in the variables shown under a nonmanipulatable category without considering the manipulatable category, it may become unsuccessful. However, when different variables are considered for evaluation, the questions about scale of measurement need to be resolved. After Anderson (1971), four types of scale may be considered:

1. *Nominal*—just naming the group or groups into one category and in this case no quantitative relationship can be made.

Table 9.2 Manipulatable and Nonmanipulatable Internal Social Process Variables	
Manipulatable Variables (Independent Variables)	**Nonmanipulatable Resultant (Dependent Variables)**
Hierarchical structure	Performance
Size	Cohesiveness
Complexity	Attitude
Formality/informality	Morale
Composition	
Type of work (social or productive)	
Work organization	
Norms	
Reinforcement	
Communication	
Leadership	
Status and roles	
Degree of autonomy	
Voluntary or involuntary membership	
Fiscal process	
Program (time spent in activity)	

2. *Ordinal*—specify a quantitative explanation among different categories or points of the scale but are limited to scale of equivalence and inequality thus of three values only: "equal," "greater than," and "less than," with no statement regarding the distance between two unequal values.
3. *Interval*—the scale is formed when the distances between any two points can be known for all the values in the scale (e.g., temperature).
4. *Ratio scale*—is one which has a characteristics of interval values as well as having absolute zero values (e.g., velocity and mass).

As the social processes are real-life phenomena, they can be measured with nominal or ordinal scales that can be designed and created on a rational basis. But in some cases, demographic variables can be scaled with ratio or interval scales. Therefore, it is important to categorize the social information. Usually, variables considered for comparison require checking the validity. According to Fairweather and Tornatzky (1977), there can be two types of validity for social information:

1. Outcome validity
2. Concept validity.

In the case of sustainability assessment of policy, concept validity seems more appropriate. This validity concerns the degree to which any assessment device measures what it purports to measure. The central question of the validity is whether the inclusion of created or chosen instruments compares the desired concept or not. Two steps may be needed to determine these forms of validity:

1. Conceptualizing different assessment areas to measure the concept.
2. Finding their central dimension.

For example, sustainability assessment of environmental role of forest land use can be assessed on the basis of existing forest area and/or reduction of water flow. Determining the central dimension (usually by factor or cluster analysis), which in this case may be called derived dimension, is also important. Each variable under consideration is then correlated with that derived dimension. If there is no central dimension, several independent measures of the concept could be used. Reliability is also another important concept needing to be determined, which usually deals with the instruments used for measuring a policy. As in this case, there may not be any mechanical instrument for measurement, more specifically a measure's reliability is the degree to which the same scores would be obtained if it were possible to repeat assessment procedures.

9.2 PARTICIPATION EVALUATION

Participation evaluation is one of the important phases of assessing the social dimension of a policy. Participation evaluation is targeted to assess the potential and performance of policy actors. However, to identify the potential of the actors, participation evaluation also includes the definition and identification of problem and evaluates the constraints and opportunities of the system. The roles and potentials of actors are clarified on the basis of constraints and opportunities the actors enjoy within the policy. Those potentials are used as a background scale for measuring the performance of the actors. Different aspects of participation evaluation are presented in the Rapid Appraisal of Agricultural Knowledge System (RAAKS) model (Fig. 9.1). Using the guideline shown in the RAAKS model, only the participatory aspect of a policy evaluation may be possible. The model can help in building up the information base, but it helps little about the physical resource base. Thus, the policy evaluations need to go beyond the model. There are some limitations

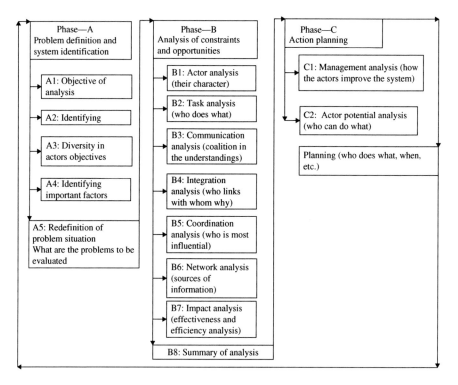

Fig. 9.1 RAAKS model. Winter (1996).

of RAAKS model to identify the roles of certain actors particularly those who influence the policy from outside the system. Therefore, process evaluation may be taken as complementary to participation evaluation.

9.3 PROCESS EVALUATION

Process evaluation deals with how the policy has been formulated and/or implemented, and thus explains how external and internal are actors involved in the policy process. The sustainability and environmental components of resource policies are now coming under the influence of many pressure groups both endogenous and exogenous. However, though there could be more than one agency involved in policy formulation, usually one agency is involved in implementation, which is important from the perspective of policy evaluation. According to Winter (1996), the models of the policy processes are:

- Formal structural model,
- Pluralistic model,

- Marxist model, and
- Corporatist model.

These policy models are not mutually exclusive. In fact more than one model could be used for policy formulation and it is not impossible that a different combination may be needed for a different society and each society may have a different process model. The institutional framework engaged in the policy formulation might have an effect on the policy output and such differences need to be considered during policy evaluation. The best way to achieve a better result is to consider policy implementation as a part of the policy process. Barrett and Fudge (1981) wrote:

> it is essential to look at implementation not solely in terms of putting policy into effect but also in terms of observing what actually happens or gets done and seeking to understand how and why.

When a policy is about resources, some negotiation is usually needed between different actors and agents, as each of them tends to maximize their benefit. The political process, by which the policy is mediated, negotiated, and modified during its formulation and legitimization, does not stop when initial policy decisions are made but continues to influence policy through the behavior of those responsible for implementation and those affected by the policy actions for protecting their own interest. This view of implementation takes away the traditional approach on formal organizational hierarchies, communication, and control mechanisms and places more emphasis on:

- the multiplicity of actors and agencies involved and the variety of linkages between them,
- their value systems, interests, relative autonomies, and power bases, and
- the interaction that takes place between the actors and the agencies.

The identification of key stages in policy implementation is a crucial procedure in evaluation and extraordinarily it is one that is neglected in official programs of evaluation. The identification of stages depends on the type and nature of the policy in consideration, but five key elements have been suggested by Winter (1996):

1. *Publicity*—potential recipients know about the policy scheme through publicity. The evaluation identifies whether there was any bias in publicity or whether some recipients heard about the scheme before the others.

2. *Advice/extension*—recipients are made involved in the scheme through advice/extension. The evaluation deals with whether sufficient advice was available to recipients or not. Whether the advice was accurate or unduly optimistic or pessimistic. Whether there was any bias in the advice so that some recipients are more or less likely to respond than others.

3. *Participation in the scheme*—identifies how the recipients are involved in the policy scheme. The evaluation delineates the administrative ease to embark on the scheme, whether any financial problem exists which may exclude the target groups and once started whether the scheme was managed and administered efficiently.

4. *Monitoring and adaptation*—identifies the attempts for solving subsequent problems. The evaluation tries to define the kind of monitoring action, how and for what objectives the monitoring action was done and whether the monitoring led to any modifications of the policy.

5. *End point*—determines whether the end point is spatially and temporally logical to the goal of the policy.

However, there is a limit to the ability to evaluate the roles of players through a process model. In the case of sustainability assessment of policies, such evaluation is generally retrospective (not progressive).

9.4 RETROSPECTIVE POLICY EVALUATION

Retrospective evaluation is different from progressive evaluation. In progressive evaluation, goals and process are given, evaluation tries to track whether the process going to achieve the goals but in retrospective evaluation status of achievement is given but there is a need to establish the goal and processes to explain why the achievement was poor or perfect. Other application of social science theory, methods, and techniques to identify and assess the processes and impacts of governmental policies and programs may remain similar. Possible steps of retrospective policy evaluation are presented in Table 9.3.

In assessing the sustainability perspectives of policies, retrospective evaluation is far and away one of the most important aspects of outlining actor relationship. Hoggwood and Gunn (1984) identified five main categories of technique for retrospective evaluation:

1. evaluating the situation before and after implementation,
2. modeling,

Table 9.3 Stages in Retrospective Policy Evaluation		
Task	Methods	Difficulties
Establish broad policy aims, i.e., policy goals	Examine contents of parliamentary acts and statutory instruments, regional directives, ministerial, government or regional statements, cabinet papers, interviews with key officials or politicians	Secrecy; conflicting goals; propaganda versus reality, ulterior motives; post hoc rationalization
Identify specific objective or desired outputs	As above with additional emphasis upon the policy refining carried out by government agencies	As above
Identify agencies and individuals responsible for implementation	As above with additional experiences upon interviews with key personnel	Multiagency projects
Specify key stages in implementation process	As above with additional emphasis upon the experiences of recipients	Post hoc relationship can be a particular problem
Assess achievement	Define criteria for success, decide on means of measurement, collect data analysis	Quality of the data, problem of other influences, or side effects
Source: *Winter (1996)*.		

3. experimental method,
4. quasi-experimental methods, and
5. retrospective cost−benefit analysis.

"Before and after implementation" might appear to offer the most straightforward approaches to assessing the achievements of a policy, but in practice they present considerable difficulties. It is rarely possible, for example, for the policy evaluator to be farsighted enough to establish the full information prior to introducing a policy scheme. Even if information is available, difficult questions may remain whether apparent policy achievements are exclusively the consequence of a particular policy.

In many ways, modeling is a refinement of "before and after" studies. Models attempt to incorporate systematically all those factors that may have influenced a policy outcome. Lack of suitable data and erroneous assumptions about causal relationships provide a formidable challenge to those wishing to evaluate policy through modeling.

As the name implies, the experimental method relies on testing a policy on a particular group of people, ideally retaining a control group with identical characteristics with the experimental group on which the policy is not applied. The experimental method in policy evaluation works best where the influence of other factors or variables can be eliminated. Thus, the method has some applications in situations where groups of people can, to some extent, be isolated. However, the heterogeneity of resources and their production base may present limited opportunity for experimentation of this kind. The monitoring of a policy retrospectively, even with the use of control groups, cannot be construed as evaluation using the experimental method, as this requires a program delivery focused entirely on evaluation considerations. However, well-designed programs of this nature using control groups (the quasi-experimental method) may offer some scope for evaluation.

Finally, the retrospective cost–benefit analysis is an extension to the methods already outlined. The prospect of attributing a financial value to the cost and benefit associated with a particular policy signifies the success of this method. It does not, however, solve some of the conceptual or analytical problems of isolating costs and benefits, establishing causal relationships, and eliminating independent variables.

In practice, most formal policy evaluations could utilize a range of methods. However, much comment on the impact and achievement of policy by politicians, policy agents, and academics may not take care of these methods. Intuition, ideology, and received opinion, all figures largely in what may pass for policy evaluation. In the same way, implementation is as political a process as policy formulation, so retrospective evaluation is inevitably a part of a continuing policy process in which political imperatives and the focus of the policy figure highly. Thereby, identification of policy focus is important for retrospective evaluation of social dimension.

9.5 EVALUATION OF POLICY FOCUS

Retrospective policy evaluation engages the nature, causes and effects of governmental decisions or policies. The purpose of policy evaluation is largely to find out the appropriate relationship between the policy focus and the policy decisions. According to Nagel (1984) evaluation of policy focus may be done by concentrating on the following ways:

1. Taking policies as givens and attempting to determine what causes them.
2. Taking social forces as givens and attempting to determine what effects they have.

3. Taking policies as given and attempting to determine what effects they have.
4. Taking effects or goals as given and attempting to determine what policies will achieve to maximize those goals.

Among the propositions for evaluating policy focus stated above, the first two steps are mainly associated with political science or political perspectives. The third is usually an evaluation research, and the fourth one is for optimizing perspective of the policy research and mostly related to sustainable development. However, policy focus evaluation cannot be totally value free since sometimes achieving policy focus may be sacrificed for achieving or maximizing certain given social values. Thus, policy evaluation often takes extra precautions to keep social or personal values from interfering with their statement of focus. These precautions can include drawing upon multiple sources and individuals for crosschecking information, making available raw data sets for secondary analysis and making assumptions more explicit. In this regard, policy evaluation may have three perspectives:

1. deducing perspectives,
2. optimizing perspectives, and
3. political perspectives.

Deducing perspectives draws conclusions concerning the effects of alternative policies from premises that have been empirically validated or intuitively accepted. Optimizing perspectives deals with deducing the conclusions to ensure what policy will maximize benefit−cost ratio taking into consideration that policies often have diminishing returns and must be analyzed in the light of economic, legal, and political constraints. The political perspective of the policy evaluation deals with the role of interest groups, government personnel, and government procedures in determining policy formation and impact. Each of the above perspectives of policy evaluation has its own merits and demerits as detailed in the following sections.

9.6 DEDUCTIVE POLICY EVALUATION

Deductive policy evaluation often refers to the more specific methods used in retrospective analysis of public policies. The main methods, however, are no different from those associated with social sciences and the scientific methods in general, except they are applied to variables and subject matters involving relations among policies, policy causes and policy effects. According to Nagel and Neef (1980), the main principles of deductive analysis are comparative modeling and deductive modeling.

9.7 COMPARATIVE MODELING

One of the scientific alternatives to deducing the effects of alternative policies involves comparing people or places who have not experienced a certain policy with those who have experienced the policy. Although this principle seems suitable for past policies, it may have a number of methodological and normative limitations:

- On a methodological level, there may not be enough places where the policies have been adopted or it may be possible that there are no places where policy has not been adopted. On the other hand, though both the groups present, it may happen that the policies have not been adopted in similar places or simultaneously.
- There may be possible that policies have been adopted by some places and not others, but the adoption has involved nonrandom selection which may tend to make the comparison meaningless.
- Another defect in the purely empirical approach is that the policy may have been adopted too recently for a long-term evaluation. In such cases, a period of several decades may need to pass and a comparison needs to be made with data before implementation of the policy and say 30 years after the implementation.

To avoid these deficiencies of the traditional cross sectional or time series analysis of policies or treatments, the principle of deductive modeling may be applied.

9.8 DEDUCTIVE MODELING

Deductive modeling considers the decisions of a policy and works them out with the implementation process. Deductive modeling may have three forms:

1. model dealing with group decision making,
2. models of bilateral decision making, and
3. models of individual decision making.

If the number of implementing agency varies, e.g., four to eight, there could be a difference in the achievement of the goals of a policy. Finding out such differences may be done by method 1. On the other hand, method 2 assumes that implementation of one decision may change the performance of other decision, and method 3 tries to illustrate the change in all other objectives due to a strict implementation of one objective of the policy. However, such a relationship is based on the assumption of a probabilistic approach that requires a large number of observations to fit in probabilistic distribution before drawing any inference. In the case of land use policies of the past, such a large number of observation is not possible; therefore, the implication of deductive modeling remains limited in this thesis.

9.9 OPTIMIZING PERSPECTIVES

The main scientific alternative to optimizing perspectives of public policies is to take policies as givens and then to determine their effects. This is often referred as impact analysis. An optimizing perspective on the other hand, takes goals as givens and attempts to determine what policies will maximize those goals. Generally, the best or optimum policies or set of policies are those which maximize the benefits minus costs subject to economic, legal, political, or other constraints. Models of optimizing alternative resource policies can be classified in terms of their involvement in the following findings:

1. An optimum policy level where doing too much or too little may be undesirable.
2. An optimum policy mix where scarce resources need to be allocated.
3. An optimum policy choice among discrete alternatives, especially under conditions of uncertainty.

In dealing with an optimum level problem 1, one needs to relate adoption costs and nonadoption cost at various policy levels. The optimum policy level is then the level of degree of adoption that maximizes the sum of the adoption costs and the nonadoption costs.

In the optimum mix problem 2, one needs to determine the relative slopes of marginal rates of return of each of the places or activities under consideration. If linear relations are present, the optimum mix is then the mix or allocation that gives the budget to the most productive places or activities subjected to the constraints. If diminishing returns are assumed, the optimum mix is the one that allocates the budget in such a way that the marginal rates of return are equalized across the places or activities so that nothing can be gained by redistributing the budget.

In optimum choice problem 3, one needs to determine the benefit and costs from each choice. The optimum choice is then the one that has the maximum benefit−cost difference, after appropriate discount. The problem of the optimizing concept with the land use policy could be that there are many issues within the environment which cannot be evaluated on standard prices, and therefore, the application of optimizing principle is limited in the field of environment. Also, the optimizing principles are applicable for policies under planning. The scope for utilizing those principles in evaluating the past policies is limited.

9.10 POLITICAL PERSPECTIVES

Political perspective is a special consideration of policy decision. If involved, the deducing perspectives and optimizing perspectives of such a decision may give negative results. Such perspectives of decision are important for international relations or to reduce social instability. Administrative influences on resource policies are mostly related to special political perspective thus are quite relevant to this study. In this principle, the policy serves some fixed objectives beyond the resource system, thus the evaluation explains what had happened to the resources (not what could happen) before placing a next interest on the resource. Thus, this principle tries to explain the post-policy status of resources and correlates them with the implementing tools used in the policy.

The above discussion reveals that there are numerous principles and tools for policy evaluation. The choice of tools for evaluation depends on the temporal status (whether the policy is planning phase, ongoing, or passed), interests, objectives, and targets of policy.

CHAPTER 10

Factors of Sustainability Assessment

10.1 INTRODUCTION

Environmental understanding about resource use is generally oriented around the concept of resource accumulation and utilization through the respective processes of harvesting or conservation. However, some authors such as Houghton (1994) acknowledge that harvesting is not always destructive if resources are selectively harvested and are not

Sustainability Assessment. DOI: http://dx.doi.org/10.1016/B978-0-12-407196-4.00010-6

particularly exposed to inappropriate human attitude; they usually recover. On the contrary, the findings of Brown et al. (1991) and Flint and Richards (1991, 1994) in forest resource studies showed that in many tropical forests the average biomass is declining by selective logging. An example was drawn from Malaysia that over the period 1972–1982 the loss of forest was 18% and the loss of total biomass was 28%. In reality, subsequent processes of logging cause more degradation than the logging itself. For example, among the predicted subsequent reasons for land use change, urbanization is considered as an important reason in recent decades (WRI, 1996). Although the expansion of urban area is not alarming in comparison to the expansion of the agricultural areas, the sprawling suburban area is displacing both agricultural and natural ecosystems by dumping and poisonous gas emissions. Therefore, resource issues for policy evaluation range from core resource operation activities to encroachment profile and involve a wide variety of agents or actors. The following sections describe some of those issues.

10.2 ACTOR AS POLICY FACTOR

Resource operations in developing countries are influenced by a range of socially visible and invisible actors from home and abroad. Identification of them in the policy may be possible by the display made by them. Boehmer-Christiansen and Skea (1991) noted that the science and politics interfaces are important fields of practices through which discursive power is exercised by the actors and which should accordingly be included in a discourse analysis of the policy process. Here, the term science–politics interface signifies that political decision making in the policy process is delimited by the scientific findings.

According to Hajer (1995), there may be three kinds of key actors operating in the field of resource policy:

1. Group of actors seeking solution of the problem who work on behalf of the government or executive agency (let us say progovernment).
2. Groups who differ with the actions of the government (antigovernment).
3. The third group is the NGO, contextual or extraregional actors who work outside the government but play an active role (neutrogovernment).

The activities of these actors may facilitate or prevent the formation of a coalition, either discourse or actor coalition, to anticipate the seriousness of an environmental crisis, and the effectiveness of existing regulations or strategies. For example, in developing society, antigovernment actors sometimes oppose a government decision, even if it is good, to make the situation politically argumentative. Such a noncoalition results in the lack of an appropriate relationship between science and policy, defies maintenance of social order, and disguises the questions of morality, responsibility, and social justice.

10.3 GLOBAL RESOURCE FACTOR

As the global influences on resources are transmitted through political actors, the action and influence of political actors determines the applied status of global influence. However, the positioning of political actors in a country depends on how the local actions are tackled. Thus, fixed at appropriate levels, different measures of government intervention are expected to secure global influence on domestic resources. Government attempts may consist of either a sustained support to domestic consumers over an extended period or a pattern of controlling international trade flow to bring equitable income distribution and overall economic welfare (FAO, 1988). Thus, resource policies aimed at satisfying domestic consumers' income in one case and essential goods supply in the other often have pronounced short-term consequences on world trade and a long-term consequence on forest health. Therefore, domestic stabilization sought by both exporters and importers of resources needs to be oriented on the buffer areas of resources. It may be thought that market protectionism may have a good effect on resource protection, but when the demand for a particular commodity grows and the market expands, unfortunately protectionism has limited effect (Amelung and Diehl, 1992).

10.4 LOCAL RESOURCE FACTORS

The above section underscores that structural and market stability of local conditions is important for sustainable resource use. However, the local actors also need to understand that if resources were lost, subsequent critical environmental conditions developed from such losses would be very expensive, and that the efficiency of the whole economy may suffer. According to some critics, these distortions in economies will constitute the main cost of resource policies in terms of their adverse effects on overall output. The main adverse effects could be:

- Barrier to industrialization
- Cost to domestic economies
  - cost to consumers—have to pay higher price
  - cost to tax payers—may need to pay more to support the program
  - cost to efficiency of the economy.

Thus, the policy evaluation of resources needs to notice how the local investment is flowing in the resource sector and how public agencies are responding to the flow of investment.

10.5 PARTICIPATION FACTOR

Peoples' participation is one of the local criteria that often determines the sustainability status of resource policy. People's involvement may be economic, social, cultural, and/or political. Participation involves at least three things (3Ps), people, process, and perspectives (objectives/resources). In case of policy participation, people may be involved either in process or as target group (perspectives) or both. Often participation in a policy process means participation in decision making which in practice may not be holistic participation if the policy does not get implemented due to lack of implementing vehicles like institutions. Therefore, when the term participation comes in, sustainability evaluation demands to know who the participants are and in what form at what stage they are involved; but they do not underscore the other components of the policy like government and institution.

Peoples' participation can be evaluated by their number and by the level of empowerment in the governance, markets, or community organizations. But as the educational profile of people in countries like Bangladesh is low, or not very comprehensive, leadership and governance capacity of the participants seldom develop other than comprehensive participation. However, it is not uncommon that the players may utilize the groups to hold their power in governance through which they can materialize their ill motives. Therefore, in developing countries it would be wise to proceed carefully to determine which discourses of participation should be appropriate.

10.6 PARTICIPATION CATALYST

Participation in developing countries is influenced by few factors, which themselves are not participatory factor but induce participation called participation catalyst. In a developing country, the catalysts often exert more influence for positive policy outcome than that of participation itself. Participation catalysts may be from the local or global influences. The most useful catalytic factors are:

- expansion of democracy (power building through participation),
- privatization (creating new avenues for participation),
- transition to market economies (dismantle centralized bureaucracy) e.g., Malaysia,
- information revolution (radio, TV, Internet), and
- NGO movement (organizing people influences the participation).

Although some of these catalytic processes are evolving in developing nations through international influence, yet in countries like Bangladesh, the majority of the people who need special attention, such as, the poor, women, and rural people are excluded from the process. In practice, if a person is poor, female, and rural, how far she is deprived from participating in the policy process is incomprehensible. In addition to these, there are certain obstacles which suppress the benign catalysts to organize these people for participation. Vested power groups utilize those predicaments for depriving people from participation. A few of those predicaments are:

- faulty legal system (do not give shelter to poor),
- bureaucratic constraints (ordinary people want to avoid),
- social norm, tradition and prejudice (e.g., regulating participation of women), and
- faulty policy (1989 land credit decision of Bangladesh facilitated 7% landlords with 37% credit).

The discussions on elements of resource policy, especially participation and catalyst, show that the discourses of policy elements function round the socioeconomic systems. Therefore, understanding the economic criteria will also be important for policy evaluation.

10.7 ECONOMIC FACTORS

Economic factors of a country may have a different pattern of influence on the policy. Macroeconomic policies, such as exchange rate, devaluation, and the level of debt servicing ratio may have an impact on the resource trade and deforestation (Barbier et al., 1992), and thus may invite catalysts of global level for policy inducement. On the other hand, microeconomic policies like micro credit and price control may influence the local catalyst to operate in the policy system. The following sections describe the influence of economic measures, specifically adhered to forest land use.

10.7.1 Influence of Macroeconomic Factors

Macroeconomic measures taken by a government are usually targeted to adjust all the policies. However, a few measures of macroeconomic policy, e.g., an overvalued exchange rate, acts as a subsidy to urban consumers on imported goods while implicitly taxing resource exports produced domestically. Real currency devaluation as frequently required by structural adjustment programs for indebted developing countries, remove existing distortions and provide incentives for greater domestic production of exportable items including primary resources. This is due to increased international price competitiveness and increased domestic demand for home-produced goods, as imported substitute goods become more expensive. Both impacts can directly encourage resource degradation through the expansion of primary production for international or domestic markets unless adjustment is made in the policy.

Generally, macroeconomic policies (i.e., fiscal and monetary policies) can affect underlying demand and supply conditions, with knock-on effects in the resource-based industry (e.g., unnecessary project aid or grant or loan is taken in a resource sector to meet the deficit of foreign reserves in the total policy). Such impacts of macroeconomic policies on the resource sector are complex to evaluate. For example, Capistrano and Kiker (1990) found a negative correlation between debt–service ratio and deforestation. In contrast, Kahn and McDonald (1990) discovered a positive relationship between tropical deforestation and public external debt. Their study also indicates a high correlation between exchange rate, devaluation, and deforestation.

However, economic policies specifically aimed at the measures to resource sector, including domestic and trade instruments (e.g., tax credits or subsidies for forest conversion, afforestation, or for wood product exports) produce a direct effect on resource degradation (Devaranjan and Lewis, 1991). Forestry is indirectly affected by economic policies which alter incentives and returns in downstream industries or related sectors, such as wood processing, construction, and agriculture (Barbier et al., 1992). Short-term concessions and poor regulatory frameworks coupled with inappropriate pricing policies often contribute excessive rent-seeking behavior in resource production (Gillis, 1990). For example, in the Philippines, if the government had been able to collect the full value of actual rent, its timber revenues would have exceeded a yearly average of US$ 250 million, nearly six times the US$ 39 million actually collected during the years 1979–1982 (Barbier et al., 1992). Instead, excess profit of US$ 4500 per hectare officially went to timber concessionaires, mill owners, and timber traders (Repetto, 1990). Similar difficulties of capturing less rent have been reported from elsewhere in other developing countries, e.g., Malawi (Hyde et al., 1991). Many of the problems may have to do with the complexity of fees and concession arrangements decided by the policy which makes enforcement and supervision of revenue collection difficult (Grut et al., 1991).

10.7.2 Influence of Microeconomic Factors
Microeconomic policies, as outlined in Hyde et al. (1991), may influence the environmental factors of primary resource management through their impacts on:

- the level of mechanical efficiency of harvests,
- the level of social efficiency of harvests (environmental externalities),
- alternative arrangements of royalty, contract, and concession and their implication for trespass, grading, and other environmental losses, and
- the level of rent distribution.

The implications of domestic microeconomic policies are also illustrated in Barbier et al. (1992) in terms of a cost curve. Lohmann (1996) reported that the optimal level of the forest operation conducted by the private concessionaires is guided by mechanical efficiency attained from the competitive price of the delivered logs, harvest volume, and short-run marginal cost. The level of such a short-run return is not optimal from a social point of view because it excludes:

- The "user costs" of short-run harvesting, that is the discounted future returns from leaving the residual stand undamaged and growing or through the avoiding of high grading and other practices that degrade the stand.

- External environmental cost of extraction (e.g., watershed degradation, downstream sedimentation, disruption to nutrient cycling, loss of natural habitats, loss of nontimber products).

The discussion on economic influence shows that economic influences create a field to guide private investments or corporations in a path desired by the policy. But investment may be public as well. If it is a public investment, the prime target of a policy may be to guide the investment to social benefit which might need to use some strategy of economic influence differently from guiding the private investment. Thus, it may be important to characterize private or public investment for policy evaluation. The following sections describe some characteristics of investments.

10.7.3 Influence of Private Investment

Privatization may not always work for environmental reasons in developing countries. Often misconception and wrong use of privatization concept appear harmful to the economy of developing countries. There are also debates that privatization supports the interest of multinationals (UNDP, 1993). Privatization should be seen as one element of a total package of stimulating private enterprises but many developing countries have taken it differently. Privatization in developing countries like Bangladesh has been politically motivated and pursued for vested interest of different interest groups or individuals rather than as coherent part of encouraging private investment. The sustainability attempt through privatization has received a mixed result in developing countries for the following reasons:

- For the wrong reason (for increasing short-term revenue rather than building competitive markets, e.g., privatization of jute and cotton mills in Bangladesh).
- In the wrong environment (privatization may bring some good result if initiated in a competitive environment rather than protective environment).
- By wrong procedure (privatization without proper circulation of information accompanied by corruption).
- For the wrong purpose (often state properties including forests are sold to cover the budget deficit or to meet the current liabilities).

- For the wrong strategy (without appropriate share distribution or without securing the jobs).
- For the wrong political consensus (political nonconcensus may result to political or social unrest in developing countries).

There are other factors such as lack of coordination and lack of monitoring that make the private motive too profit seeking, and thus environmental benefits through microeconomic adjustment may be lost. In these cases, public investment plays a key role to keep the private investment competitive and to keep the trend of coordination and monitoring perfect.

10.7.4 Influence of Public Investment
Public investment in the nonstrategic resource sector may be targeted to gear the dynamics of competitiveness of private investment and to increase social benefits through monitoring and coordination. Privatization may create sluggishness if public investment creates a monopoly on the resource sector. However, public investment may influence sustainability in developing countries due to following situations:

- too large a public sector,
- obscuring the relationship between private and public sector (e.g., budget leakage),
- inefficient public sector (affected by strike).

However, if public investment is made for the benefit of people, it certainly will bring benefit. There are examples where public investment has done good, mainly stimulated by the participation of people. Other than this, there is every chance that public investment will be a reflection of colonial motive.

10.7.5 Influence of Economic Incentives
Economic incentives are treated as an instrument to increase the peoples' participation in the public sector. Incentives may also be used to increase the efficiency and competitiveness of the private sector. But if misplaced, such as for giving benefits to cronies, the economic incentives may be the potential reasons for resource degradation. Amelung (1991) identified that the major shift in forest land use change and forest degradation in tropical countries were due to economic incentives in the forms of concessions, subsidies, and cash crop conversion.

Binswanger (1989) and Mahar (1989a) highlighted that the role of subsidies and tax breaks, particularly for cattle ranching, was one of the reasons for land clearing in the Amazon. Schneider et al. (1990) and Reis and Marguilis (1991) emphasized that incentives on agricultural rents had encouraged small-scale frontier settlement in the Amazon region. Indeed, the agricultural sector showed the largest share (over 80% on average) of total deforestation in all tropical forest countries (Barbier et al., 1992). The direct impact of forest activities in deforestation was minimal (>10%). This was partly because most commercial logging used to be done by selective logging. Amelung (1991) has shown the details of two alternative measures of forest degradation, biomass reduction and forest modification.

Although the economic influence on resource operations is clear to the governments and their advisors, in most cases, the attitude of people and governments of developing countries toward resources cannot be explained by economic principles or does not reflect in the economy of the country. Perhaps there are certain other factors, which are not explainable by economics alone. The influence of policy actions and the nature of administration together may ruin the resources of developing countries. Thus, it is important to evaluate administrative criteria pursued in policy implementation.

10.8 ADMINISTRATIVE FACTOR

Administrative factors of a country can be perceived as the role of organizational, institutional, and legal arrangements. Institutional and legal arrangements governing land tenure and transactions can have significant effects on resource use. For example, in some countries, property laws establish deforestation as a prerequisite of formal claim over the land for those settling in forested areas (Mahar, 1989b; Pearce et al., 1990). Large areas of old growth forests and economic forest zones in the Amazon have been lost by such exploitations (Grainger, 1993). Moreover, many governments are unable to effectively manage the public forest estate, resulting in illegal encroachment of logging. Short concession periods for logging operations on public land reduced the incentives for reforestation. Stumpage and license fees were frequently set at very low rates, which fail to reflect the scarcity value of standing timber. The case for fisheries resources of Bangladesh is similar. All of these are administrative factors have happened due to weak institutional arrangement. The following sections show the limitations of different aspects of administration.

10.8.1 Right and Tenure

Right and tenure are administrative criteria with which access and admission to the resources are controlled. In the absence of right and tenure, a resource area essentially becomes an "open access" resource from which no one can be excluded. Under those circumstances, individuals may try to maximize their short-term benefit in a competitive fashion, leading to a quick exhaustion of resources. Failure to design appropriate concession arrangement for public resource areas and insecurity of ownership of new resource development (e.g., plantation and fish culture ponds) can create similar conditions of open access situations. That is why private individuals and their concerns usually make harvesting decisions based on short-term profit maximizing decision and may have little regard for the potential for greater future returns from the timber stand.

10.8.2 Decentralization

Decentralization may be one of the best options for undertaking resource operations and participation efficiently. In developing countries, like Bangladesh, where the history was centralization of power, decentralization of power both waxed and waned at different periods, particularly when governments changed. In long-term planning, as in forestry, lack of decentralization or frequent change of decentralization of power creates barriers to the sustainable management of resources. Although there are certain institutions for decentralization, e.g., Local Government (Union Council), Upazilla Parishad, in Bangladesh, villages in Indonesia (UNDP, 1993) and "Barangay" in the Philippines (de Guzman and Padilla, 1985), the institutions were more inclined to economic development than to environmental sustainability. Moreover, the structure and power of those local institutions have been changed frequently in Bangladesh with the changes of Government (Hye, 1985). Even the centralization concept in developing countries only works in limited cases like economic disbursement. On the sectors like social security, there is a weak centralized control; thereby, the centralization itself does not get completed. Different forms of decentralization have been shown in Fig. 10.1.

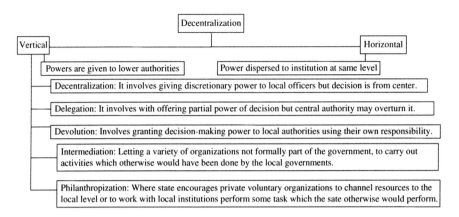

Fig. 10.1 Characteristics of decentralization. UNDP (1993) and Uphoff (1985).

Not all forms of decentralization would be useful for resource sustainability of a country. The utility of particular form of decentralization depends on the consideration of size of the country, population, and other sociopolitical factors (UNDP, 1993). For example, densely populated countries require more layers of governance than sparsely populated countries. In the case of highly populated countries, lower layer governance should have the equivalent power of the lowest layer of governance in sparse populations. However, quantities estimated from decentralization should be treated with care and need to be complemented by a broader knowledge and understanding of concerned countries. The authority and emphasis of decentralization may be expressed by the following terms:

- The expenditure decentralization ratio—the percentage of total central expenditure expensed by local authority of the resource department.
- Modified decentralization ratio—assuming that certain expenditure cannot be decentralized, e.g., staff salary, what proportion of the rest is spent by the local authority.
- Revenue decentralization ratio—it is the percentage of local authority revenue to the total central revenue. The larger ratio shows higher revenue decentralization. In Bangladesh, almost all forest revenue is decentralized but controlled by the central power.

- Financial autonomy ratio—it gives an indication of local authorities' independence from the central authority. It is the percentage of local authority revenue to the total expenditure.
- Employment decentralization ratio—local authority employees as a percent total employees of the department.

Although the reflection of decentralization in total governance could have an effect on the decentralization of a resource management system, the decentralization pattern of one country may not be the same and similarly useful for other countries. For example, the Indian mechanism of decentralization adopted the Franco mechanism, unlike the British or American system, in which several facets of deconcentration and decentralization combined (Muttalib, 1985), whereas in the Philippines it is a democratic decentralization, merely an American system (de Guzman and Padilla, 1985). Thus, the financial position of local government and its relationship with central government shape the pattern and sometimes power of local governments (Muttalib, 1985).

10.8.3 Accessibility

Sustainability of a policy depends on accessibility in formulation and control of resource accessibility through policy implementation. Accessibility during policy formulation establishes democracy and social participation ensuring a fair assessment of the needs of society and expectation of policy. Policy sustainability thereby can be arranged by negotiating between needs and expectation at the very beginning of policy.

Policy implementation on the other hand controls the accessibility of resources. Accessibility speeds up the transformation of resources. This happens when the access cannot be controlled administratively. Thus, a road network, is recognized as a useful tool for forest development in developed countries. In developing countries and particularly in highly populated countries, road development beyond a controllable limit causes resource degradation. Reis and Margaulis (1991) have found a correlation between the increase in road construction and an increase in the rate of deforestation in the Amazon area. Regarding the biodiversity loss as well, Reid and Miller (1989) claimed that the decline in rich biodiversity (50–90% of world biodiversity) of tropical forests gets affected more by road construction than the selective logging. Thus, for developing countries the road network can be an indicator of forest sustainability.

10.9 MARKET INFLUENCE

Market behavior of a policy is usually guided by the economic and trade balance relationship. Environmental consequences are seldom considered in the market analysis of resources. Thus, if the domestic/ microeconomic policy efforts to manage resources do not go right, the sustainability issues are likely to be affected invisibly and negatively by market behavior. To face such negative impacts, domestic policy on the resource sector seeks solution mainly in the following ways:

- providing protection (subsidy),
- pricing products (taxation), and
- negotiation.

The domestic response to market influence may include one or all of the above measures depending on the requirement of the country. Unsustainability triggers when such measures benefit a group of people only. Eventually, those people may become almost controllers of resources or players to the government.

In a usual situation, governments sacrifice resources for protecting the job, for earning hard cash or for other state emergencies. Governments may have to accept a lower global price or may need to protect the exporters to compete in international markets (Grainger, 1993). If market prices fail to account for indirect use values and future nonuse values (Rees, 1990), they might be lost or degraded by direct consumption of forest produce. When forest users fail to appropriate environmental values, they will tend to ignore them. Logging companies, for example, may neglect the impact of logging on wildlife and tourism. When such external costs are consistently ignored, prevailing market prices of those goods tend to fall below the social optimal limit leading to excessive exploitation (Eckersley, 1993).

The market prices of the traded resource, products, and services may also be distorted by other public policies, including economic policies, public investments, and institutional arrangements (Wilson and Bryant, 1997). Barbier et al. (1992) observed that it is often difficult to disentangle the linkages between domestic macroeconomic, sectoral, and trade policies as they affect forestry. Economic policy interventions at various levels can alter the profitability of the forest-based activities vis à vis other domestic sectors and thus their market competitiveness. Capistrano (1990) and Capistrano and Kiker (1990) explored the international and domestic macroeconomic factors on tropical deforestation. They revealed that the export value of the tropical wood was a major factor explaining the depletion of closed broad leaf forests between the years 1967–1971. Barbier et al. (1992), Grainger (1996a), and Johnson (1999) revealed that in Thailand and the Philippines, once net exporters of timber are now net importers, due to such failure of pricing forest products.

Grainger (1993) suggested that the import and export policies influencing resource trade flows can themselves be a source of major environmental impacts in the resource sector. Protection of resources and resource-based industries, whether explicit or implicit, can encourage inefficient expansion of domestic industries that leads to excessive resource depletion. Producer countries that are unable to penetrate protected international markets may be discouraged from developing value-added processing, even though a comparative cost advantage may exist in these countries for such industries. It is conceivable that the loss in value-added process leads to a higher rate of resource exploitation in order to increase earning from raw material and semi-processed exports. However, overexpansion of processing capacity and inefficient operations could result in greater resource loss.

10.10 HISTORICAL FACTOR

The sustainability of a policy increases if the policy is formulated on the basis of evidence. Consideration of historical issues is important for establishing policy evidences because of social criteria like poverty, social structure, and tradition; administrative criteria like organizations and institutions are historical elements of the society. Historical constructs of environmental crises have largely influenced the present policies of most developing countries (Bryant, 1997; Guha, 1989; Jewitt, 1995). Bryant (1998) cited an example on such historical construct that the introduction of scientific forest management in Asia by the British rule was not intended to benefit the forest but to promote the long-term commercial timber production of special categories like teak and deodar. One of the targets of the scientific forest management was the production of key species by eliminating forests of other species and to remove the resistance of local people or to restrict the alternative indigenous practices like shifting cultivation (Bryant, 1994; Gadgil and

Guha, 1992; Jewitt, 1995; Peluso, 1992). The legacy of scientific forestry in India subsequently brought many changes in forest land use and gave a shape to the forest management and administration different from local appreciation. In doing so, the organization of forestry staff was fabricated in a form of policing power, which is still continuing, and resulted in the separation of people from the forest administration. Thereby, a discourse of forestry practices developed in which appropriate forest use was defined largely in terms of commercial extraction, which was asserted to be scientific and also financially remunerative to the state. As a consequence, the local practices disintegrated, were marginalized, and in some cases were criminalized (Bryant, 1996). Thus, historical issues merit evaluation for the sustainability of forest resources.

10.11 OTHER FACTORS

From the above section, it is understandable that the roots of many factors of policy climate are in the past policy processes of the society. Houghton (1994) indicated a particular period of the past when major forest degradation took place,

> though the settled agriculture has been practised by mankind from more than 10000 years ago, major changes in the land use had happened in the last 100 years, and particularly for developing countries in the last 50 years.

Therefore, it can be assumed that there are some other factors blended with the past processes of 50–100 years bringing the present change in forest land use. The vital component of sustainability research in developing countries is usually directed to the population explosion and its consequences on forest resources (Bryant, 1998). This form of research partly reflects the nonusefulness of neo-Malthusian notions in developing countries, because of the limitation of such proposition to explain the circumstances of the environmental crisis. During the 1980s, the discussion was prevalent and was reflected in the work of Buchanan (1973), Darden (1975), and Lowe and Worboys (1978) asserting that technological innovation is not effective in developing countries. The study of the people–environment relationship has been taken farther in research influenced by household studies and ecofeminism to examine how power relation within the household influenced the control of land, natural resources, labor, and capital. Subsequently, policy can start to draw social structuralism and social discourse to integrate the power and knowledge for a successful incorporation of environment and sustainability within the social policy (Bhaba, 1994; Peet and Watts, 1996).

The issues of sustainability assessment of policy thereby look at the magnitude of the following attributes to see how they have been established in policy formulation:

- defining a clear overreaching objective,
- mapping the policy context,
- identification of key stakeholders and addressing issues related to them,
- identification of behavioral changes,
- development of strategy,
- analyzing internal capacity to influence desired change, and
- establishing a monitoring and learning ability.

If we look at the above attributes, we see that each of them may require multiple tools (not discussed here) and has the potential to induce changed over the period of policy cycle. Sustainability of policy thereby remains in the tools considered for ensuring those attributes and options are kept within the policy to deal with the changes in a beneficial manner which is usually known as policy flexibility.

CHAPTER *11*

Tools for Sustainability Assessment

11.1 INTRODUCTION

The tools for sustainability assessment could be numerous; entailing social, environmental, economic, and resource issues related to policy. However, optimum choice of tools for sustainability assessment is possible if enough information is available about the policy variables. Gathering and analyzing information not only helps focus attention on important issues but also helps to limit the information required while broadening the question to be addressed. According to NCEDR (1998), the tools for aiding sustainability decisions in environmental policies are:

- identifying environmental values,
- characterizing the environmental setting,
- characterizing the social, political, and economic settings,
- characterizing the legal setting,
- integrating information,
- forecasting,
- assessing options, and
- conducting post-decision assessment.

Assessment tools are those which can explain the role of the above decision aids from the outcome of the decision; thus, in this case, they

Sustainability Assessment. DOI: http://dx.doi.org/10.1016/B978-0-12-407196-4.00011-8

mostly deal with the suitability of the decision aids. Use of tools helps in identifying options in the context of decision by identifying, specifying, and assessing relevant issues and information. However, choosing tools needs some balancing because speedy evaluation needs to be balanced against the objectives of broad participation, or the objectives of full detailed information need to be weighed against the objective of containing the costs of the assessment process. In the specific case of forest land use, the outcome of environmental values of policy decision is the quick process, but people's attitudes and preferences toward the natural resources resulting from such a decision can be incorporated for detailed policy evaluation. In this instance, questions may be asked about the tools for determining environmental values of a decision. Some of such valuation tools are:

- economic markets,
- ecological relationships,
- expressed-preference surveys, and
- small group elicitation.

Economic market analysis is a cost-based evaluation and ecological relationship can be considered as a moral-based evaluation. However, it is difficult to separate the moral, ethical, and economic elements incorporated in the concept of environmental values. Individuals hold many values such as, scientific, cultural, esthetic, religious, recreational, economic, and other environmental values. Environmental values comprise attitudes and preferences toward all aspects of natural resources. These values are derived from direct and indirect use of natural resources (Gregory, 1998). Therefore, the analysis of value is mostly subjective, up to the requirement of the policy and the society.

The consideration of policy factors depends on the people who are engaged in the evaluation. For example, to the technical experts, policy evaluation is a profession; they will try to choose more accurate tools, thus may deal with a factor at a time. To the administrative consumer, policy evaluation is an aid to decision making thus more attention may be given to legal factors. Politicians may be interested in political and social factors, and to the ordinary people, policy evaluation can be a bureaucratic waste because they are interested on the personal economic interest level only. Serving all the groups by policy evaluation may be difficult. But there are certain models (tools of policy evaluation) which make the policy evaluation simple and acceptable to all the groups. According to Byrd (1980), the characteristics of such tools are:

The model should be simple, robust, and easy to control. The model should not give absurd answers.

The model should be adaptive. As policy decisions are not static, policy evaluation should be adaptive to the new situations.

The model should be complete on important issues. There are issues in policy where measuring parameter is difficult. Those issues need to be accommodated within the policy analysis so that subjective consideration of those issues can be avoided.

The model should be easy to communicate. This character has relationship with the characteristics in number 1 and 2.

As a result of these need assessments in policy evaluation, a policy analyst is supposed to be guided to a new form of assessment tools which should be more simple, less sophisticated, and highly oriented to user groups. The goals of the models should be insight not the numbers. This type of modification in the policy evaluation model can be termed as "humanization" of policy evaluation.

There are different tools for evaluation of social aspects of policy. The tools are used to assess how different policy factors have been considered to meet social aspirations. There are no exhaustive lists of such tools. The following list gives some of those tools that can be used for evaluating different factors of policy. Each of the tools merits different strategies and statistical approaches for assessment of policy factors, the detailed discussion of which are beyond the scope of this book. We are mentioning some of those tools applicable for the specific evaluation of policy factors:

1. Mapping policy context: Mapping the policy context means identifying key factors that may influence the policy process. The tools are usually arranged to answer the following questions:
   (a) How could a policy influence the political context?
   (b) How do the policy makers perceive the problem?
   (c) Would there be any political interest in changing the policy?
   (d) Is there enough evidence to take specific policy prescription?
   (e) Who are the key organizations and individuals with access to policy makers?

(f) What are the agenda where external actors/donors are involved?

(g) Are there existing networks to use?

The above questions have queries on both factors and actors of policy. For answer those questions, the following tools can be used:

(a) policy driver assessment,
(b) power analysis,
(c) strengths and weaknesses analysis,
(d) opportunity analysis,
(e) threats analysis,
(f) influence mapping, and
(g) force field analysis.

2. Identifying key stakeholders: Identifying the key influential stakeholders and target audiences involves determining what are their positions and interests in relation to the policy objective. Some can be very interested and aligned and can be considered natural allies for change. Other can be interested, though not yet aligned, and can yet be brought into the fold of reformers so they do not present obstacles. The possible tools for identifying stakeholders are:

(a) alignment, interest, and influence matrix analysis,
(b) stakeholder analysis,
(c) influence mapping,
(d) social network analysis, and
(e) force field analysis.

3. Identifying desired behavioral changes: Identifying desired behavioral changes entails describing precisely the current behavior and the behavior that is needed, if the key influential stakeholders are to contribute to achieve sustainability of the desired policy. It also calls for short- and medium-term step-changes that can be monitored to ensure that the priority of policy is moving in the right direction:

(a) progress markers,
(b) opportunities and threats timeline, and
(c) policy objectives analysis.

4. Strategy development: Strategy development entails spelling out milestone changes in the policy regime. The probable tools that can be used are:

(a) force field analysis,
(b) communication analysis for strategies,
(c) analysis of advocacy campaigns,

(d) network functions analysis,

(e) analysis of structural strategy, and

(f) analysis of research strategies.

5. Internal capacity of policy: Internal capacity of policy entails the analysis of motivation of policy team and their competencies to operationalize a strategy, In other words, the team must have the set of systems, processes, and skills that can help the policy to success in all conditions. The tools for gathering information toward evaluation of internal capability are:

(a) policy entrepreneur evaluation,

(b) strengths, weaknesses, opportunities, and threats analysis, and

(c) internal performance framework analysis.

6. Establishing monitoring and learning frameworks: To develop a monitoring and learning system not only to track progress, make necessary adjustments, and assess the effectiveness of the approach, but also to ensure sustainability. The tools that are applicable for evaluation monitoring frameworks are:

(a) logical framework analysis,

(b) outcome mapping,

(c) journals or impact log analysis, and

(d) internal monitoring tools analysis.

According to Nagel (1984), the tools can be adapted and modified depending on the application of principles of different sciences. For example, application of polluter pay tax is an economic tool and subsidy assessment is a political tool for policy evaluation. The criterion for easy communication of policy tools is satisfied by a very simple model of tools known as resource indicators. The following section summarizes resource indicators of policy evaluation.

11.2 INDICATORS FOR EVALUATING RESOURCE DIMENSION

A policy indicator, in combination with the indicators of resource properties, should represent the dynamism of the resource sector, i.e., the flow of the resources should be reflected in the indicators. For example, in the field of forestry, the information on stock is usually provided by established terms like the area of the forest. But the proportion of forest to the mainland or the distribution requirements need to be constructed. If such a construction is made for a time-series analysis, a flow of information needs to be represented by the indicators. Often it is possible to calculate from the basic flow of statistics or "primary" indicators to "secondary" indicators which cannot be directly observed but which may reveal more about the nature of the social system than the primary statistics. A "secondary" statistic usually uses a theoretical model as a basis for transformation.

The choice of indicator is governed by the purpose for which it is going to be used. The indicator chosen should be useful in recording the state and progress of a particular policy or nation; secondly, it should be useful in formulating policy. Culyer et al. (1972) illustrated that policy making requires three different kinds of indicators, each serving a different function and each complementary to the others, but none is sufficient alone for policy making. The three requirements are:

1. A measure of the output of policies.
2. A measure of deriving the social valuation placed upon different outputs.
3. A measure of technical possibility of increasing output.

Together, adequate information on each of these measures is sufficient to evaluate a policy. Such cumulated information should provide:

- The units in which the policy objectives are to be defined.
- Values/increments in each objective in terms of social worth.
- Specification of physical possibility. For example, how much of one good must necessarily be sacrificed in order to obtain more of another.

Corresponding to each of these functions, three kinds of social indicator are required in the field of forest land use:

1. Measure of State of Resource (e.g., forest land use (SOR indicators)).
2. Measure of the Need for Resource (NFR indicators).
3. Measure of the effectiveness of Activities on Resource (EAR).

Each of these indicators can be used at an aggregate level for explaining and for comparing policies, but there could be some ambiguity particularly in conceptualizing the kind of indicators applicable for different countries. However, some effort may be necessary to determine the indicator that will be more practicable in corresponding to what is ideally required for a policy evaluation.

11.2.1 SOR Indicators

SOR indicators are not specifically related to any specific input for promoting resources rather, they are measures of the phenomena that inputs are supposed to affect. SOR indicators are desired because they enter as arguments in a social welfare function for their maximization. The SOR indicators can be used at various stages of aggregation including the national and international data on the same resource. Comparison of SOR at less aggregate level is possibly useful for the comparison of policy performance in between different geographical regions. The sorts of practical use to which they may be used include in a quantitative expression of the choices that are available to decision makers, making a better international comparison or testing to see whether the resource processes of a particular country in general are conducive to environmental requirements. SOR indicators may be concerned with:

1. The total amount of resource, extent of area of resource, area as a percent of total area.
2. Resource distribution (may be presented by the following formula):

$$\left[\frac{\text{Boundary length area containing resources}}{\text{State boundary length}}\%\right] \bigg/ \left[\frac{\text{Resource area}}{\text{state land area}}\%\right]$$

3. Resource quality indicator, e.g., average biodiversity indicator for a forest.

The basic assumption in the formula of distribution indicator is that the perimeter is given, a circle contains the maximum area with minimum perimeter. For a quality forest land use, all the three indicators should be significantly positive.

11.2.2 NFR Indicators

NFR indicators can commonly be expressed in terms of a target level that a particular SOR indicator should take up. Alternatively, they may be expressed as the difference between a target and the current level of SOR indicator. The NFR indicators thus can be expressed in terms of the SOR indicators. This approach appears to have some difficulties, since it is not clear how to decide the target. With the globalization of resources and environmental concepts, targets may be fixed at a national or an international desired level but the NFR indicator does not describe whether the target is regardless of the cost or not.

11.2.3 Effectiveness Indicators

An effectiveness indicator measures the increase in the SOR indicator to be expected from a change in resource status/health affecting inputs such as fertilization, irrigation, seeding, and/or hoeing. The objectives of such indicators would be, for example, to demonstrate how, by varying one such input, the SOR indicator responds during a different time period or to show how different inputs may substitute for one another in promoting a given SOR or change in SOR. Effectiveness indicators are a sort of input–output relationship—use of such indicator is difficult in the case of resource because the production system takes a long time to give the output and there might be multiple factors behind the growth.

11.2.4 Comparing Indicators of Resources

SOR indicators, being free of any input content, should be distinguished from the NFR indicators. Cohen (1968) has stated that:

> there is a need for statistics which indicate clearly and precisely present conditions in the society including the magnitudes of existing social problems and their rate of changes.

If the statistics fail to make this distinction clear, the magnitude of social problems can only be properly measured by taking into account society's preferences and the opportunity and cost of satisfying them. That is why the NFR indicators should not be subsumed under SOR indicators. Policy values may enter into indicators either "directly" (e.g., NFR indicator valuation) and "indirectly" (e.g., relative value of change in SOR indicator to the change in some social indicators like education or income). They signify that though the analysis is about the environmental component of resource policies, its relevance to other social factors can hardly be avoided. Therefore, characterization of some indicators, such as evaluative as opposed to informative, or their scientific use as opposed to general use, cannot be considered very helpful in policy evaluation. If the society's preference remains constant, the value of a given change in an SOR indicator will depend on the level of the indicator as well as on the level of other social indicators. Even if a high value is placed on a given change in SOR indicators, it would be necessary to identify a NFR indicator along the line before the change in the SOR indicator could be adopted as goal. Such a valuation strategy is useful when the policy evaluation uses the cost–benefit approach.

11.2.5 Explanatory Variables

Explanatory variables of environmental sustainability go with feasibility and rationalization discourses—which may depend on the resources as well as the society associated with the sustainability movement. According to Hajer (1995), the feasibility includes three criteria:

1. effectivity,
2. efficiency, and
3. equability.

In sustainability issues, these terms imply a global application. This is also a difference of radical environmentalism and sustainability. Thus resource continuity, an implied purpose of sustainability, needs to agree to competitive functions with other components of the global system. Therefore, the sustainability effort of a policy would best be effective if concerted efforts can be given on regional or global scales. However, feasibility of environment is multidimensional; as much as it is benign it is feasible.

Limits can be placed on the environment on the basis of social acceptability and/or cost–benefit suitability. Social acceptability of a policy not only depends on the policy itself but also on the institution. A benign policy from an autocratic institution may not be acceptable to the society. This imposition does not necessarily mean that the changes in the institutional practices need to adopt an intermediary policy between environment and economic growth but a rational approach may be adopted which is not biased too much against the opinion of the society. No doubt sustainability is an expert-laden concept, but if only expert discourses are applied in the decision-making process, the decision may become unsustainable/unacceptable to the society. To say in short, environmental sustainability fosters a public domain where social realities, social phenomena, social preferences, and social aspirations are to be respected in decision making. Therefore, the institution involved in sustainability issues is expected to be well represented by people of the society. In this respect, it is worth mentioning that social variables like education, income, employment, population, culture, and other relevant sociocultural variables signify how well the people could represent the institutions, therefore, these variables are important to analysis of environmental sustainability within a policy. Pillai (1996) suggested that in the case of resource policy evaluation of following few explanatory variables relevant to the environment could be important:

1. Population—has a negative effect on the resource.
2. Development—has an effect on natural resource, the effect can be both negative and positive, depending on the nature of development.
3. Process of transformation is again related to population and development. In certain cases, transformation could be more devastating to resource depletion than population size. Because it takes resources directly, it causes an increase in consumption, and pollution to the environment (e.g., urbanization and forest).
4. Process of expansion is another major cause, which is also related to the population increase and development. For example, often cash cropping, not for the local consumption but for export, causes the degradation of the forest.
5. Affluence: "Affluence is polluting," the debate is seen in the agenda of development politics. Affluence is consumptive as well.

A school of policy specialists agrees that the effect of affluence on environment is mediated by a large number of political and social variables (Tobey, 1989). In addition, they argue that per capita increase in income has two influences:

1. It decreases poverty related pollution.
2. At high levels of economic growth and development policies are likely to emerge which protect the environment from degradation.

The indicator of affluence is usually taken as GNP per capita and nature of development. In terms of resource use, the nature of development is usually reflected by the characteristics of urbanization. Urbanization factors can be considered by two variables:

1. Urban population as a percentage of total population.
2. Percentage of total population residing in standard cities, with at least 750,000 people as per World Resources Institute (WRI) standard.

For an indication of development, indicator of technology can also be considered which is expressed by ENRINTY = [(ENR per Cap)/(GNP per Cap)], where ENR means energy requirement. On the other hand, energy required for one unit of GNP per capita achievement could be treated as the technological efficiency. An increase in the amount of energy that is required for producing a unit per capita income indicates a decrease in the level of technological efficiency. From the above discussion, it reveals that variables like cultural variation are important for sustainability assessment because impacts of population, development, and growth are accentuated by cultural change.

11.2.6 Tools for Assessing Human Dimension

Human dimension is the main concern in resource and environmental sustainability. Whatever may be the causes of unsustainability, they are directly or indirectly due to the human dimension. However, evaluating the human dimension is not a straightforward issue in policy evaluation. Sustainability changes in resource and environment have been described by Turner et al. (1990) in two classes:

1. Systemic changes resulted from biogeochemical cycles which operate globally.
2. Cumulative changes result in changes in the surface of the earth independent of geochemical flows.

Where the human dimension is concerned, the latter class of environmental change is important. However, the cumulative changes also affect the systemic change of biogeochemical cycles thus both classes can hardly be disintegrated. These classes of change appear as global concern when the magnitude of flow of the cycle is affected significantly or if in a particular state the accumulation of surface impact (e.g., deforestation) reaches a large magnitude, large enough to become a global problem. If the particular aspect of land use or deforestation is considered, the factors of cumulative change may be linked with the six clusters of factors: population, technology, affluence or poverty, political economy, political structure, and beliefs or attitudes (Kummer and Turner 1994; Meyer and Turner, 1992; Stern et al., 1992). The first three clusters are associated with Ehrlich/Holdern formula: $I = PAT$ (Impact = Population, Affluence, and Technology). In the statistical analysis of national to global level of cover change studies, the effects of first three clusters have been found significant (Kummer and Turner, 1994). The last three clusters in fact are the resultant imposition of first three but they are important as separate clusters because political economy and political structure also independently affect the attitude of people. Thus, if the example of forest cover change is taken, peoples' attitude toward forests is influenced by the other five clusters.

Regional and local case studies may demonstrate a great variety in the combination of human forces and attitudes contributing to resource change. These forces usually include, government policy, changing rules of resource allocation and other variable associated with political economy, political structure, and traditional belief (Allen and Barness, 1985; Blaikie and Brookfield, 1987; Flint and Richards, 1991; Kasperson et al., 1994; Kummer and Turner, 1994). Unfortunately, human attitudes do not lend themselves well to statistical analyses, due to the lack of standardized worldwide data, multiplicity of definition, and measures of proposed significant variables. Thus to be more robust, models should be more sensitive to regional resource dynamics (Turner et al., 1993). Kummer and Turner (1994) reported that sensitivity could be achieved only by delineating subglobal spatial units of common land use and land-cover dynamics.

Regional modeling of cover change may be justified under the structural concept of globalization, political economy, or political organization. In globalization, only a few are powerful and dictate the flow of benefits. Political economy ended up applying to only a few general classes like the underdeveloped, developing, and developed. Political organizations are also very few like democratic, socialist, or autocratic societies. But within those clusters the traditional belief of people varies immensely. Thus, the regional modeling of resource change may not be appropriate to determine the changing pattern of people's attitude.

Ecological footprint (EF) and the Environmental Sustainability Index (ESI) are two distinct, independent attempts specifically aimed at assessing sustainability related to the human dimension (Samuel-Johnson and Esty, 2001). The EF is a composite index involving many variables, which focus on the nature and productivity of land resources, variability of human consumption patterns, and the energy accounting of each nation's international trade. The land variables focus on area extent, biological productivity, and waste absorption capacity. The consumption variables characterize and account for the differing ecological impact of human consumption throughout the nations of the world. Finally, the EF index tries to capture the separation of production and consumption by looking at the import and export goods of each nation to see who is actually consuming the energy associated with manufacturing, agriculture, etc. (Chisolm, 1990).

ESI attempts to develop a transparent, interactive process that draws on rigorous statistical, environmental, and analytic expertise to quantify human impact on environmental sustainability. The ESI is derived by averaging five key components:

1. environmental systems,
2. reducing stresses,
3. reducing human vulnerability,
4. social and institutional capacity, and
5. global stewardship.

The "Indicator" variables that constitute the five key components are themselves derived from many specifically measurable and nationally aggregate variables chosen from complexes of resource, environment, and human attributes. Examples of a few of the fundamental variables are:

- urban SO_2 concentration,
- total fertility rate,
- scientific and technical articles per million of population, and
- number of memberships in international environmental organizations.

The significance of regional variation of human factors on resource degradation has been explained by Kummer and Turner (1994) drawing example from the forests of the Philippines. The quantitative assessment between human forces and land use change was possible. The factors of deforestation can be explored quantitatively by model as well as linking the variables (Bilsborrow and Okoth-Ogendo, 1992). Two postulates may be considered as immediate causes of deforestation: logging and agricultural expansion. In the explanation, deforestation can also be considered as two-step process, conversion of primary to secondary forest by logging and then removal of secondary forest by the expansion of agriculture, largely small holder subsistence cultivation. A panel analysis of province-level data of the Philippines was used by Kummer and Turner (1994) to explore the model (panel analysis is similar to regression analysis except observations are taken from relatively few points in time). Other factors, such as corruption, may also be important in the quantitative analysis of deforestation and have repeatedly been identified as a factor of forest removal in the Philippines (Kummer, 1992; Porter and Ganapin, 1988). Unfortunately, such factors like corruption and poverty cannot be explored quantitatively because of insufficient data.

CHAPTER 12

Problems in Sustainability Assessment

12.1 INTRODUCTION

Policies are often so complex, far-sighted, and involve so many actors that sustainability assessment cannot be considered as a straightforward issue. While preimplementation assessment of policies is likely to be predisposed and prejudiced; postimplementation assessment of policies may raise conflicts among actors involve in implementation or stakeholders of policies. Despite the universal appeal of sustainability as a general principle of development, implementation of sustainability assessment of policies would be daunting perhaps because of additional resource requirement and concern of bottom line. In considering the policy discourses, distinctions, or legitimate modes of operation often have meaning to the extent that they become inapplicable but the roles

Sustainability Assessment. DOI: http://dx.doi.org/10.1016/B978-0-12-407196-4.00012-X

and conventions that constitute the social order have to be constantly reproduced and reconfirmed in an actual situation, documents, or debates. As a consequence, the power structure of society should be studied and should be free from bias. Otherwise there would be evaluation difficulties in identifying the source of problems—whether from policies or from actors.

As such, there could be several important barriers to effective policy evaluation that can impede determining what actually occurred as a result of government action and more specifically determining the level of performance of a public policy of program. It is equally important for both policy evaluators and the consumers of policy evaluation to understand the basic nature of these difficulties. Specifically, anyone who engaged in policy evaluation must appreciate the degree of bias and the unintentional or intentional confusion that can occur as a result of difference between what is being measured during a policy evaluation and what policy makers thought the policy should achieve. All too often evaluators, administrators, and the public focus on measurement statistics that are easy to obtain but have no real relationship between what has been accomplished compared to what the original intent of the public policy was. The barriers may originate from the confusion of following areas:

- Identification of hidden goals and strategic changes in goals.
- Selecting variables that could measure the intended component of policy.
- Errors in measuring, estimating, and weighing specific issues.
- Political and players role.

This has an interesting consequence in identifying problems of policy evaluation through which eventually interpersonal communication may become more clear and relevant. It becomes imperative to examine the appropriate idea of the reality or of the status quo as something that is upheld by key actors through discourse. Likewise, it becomes essential to investigate the specific way in which the additional forces seek to challenge the constructs of a policy. Policy evaluation, therefore, is not only the evaluation of subject position but also of structural positioning. Thus, there could be numerous problems in policy evaluation mainly involving the differences of written information and policy implementation. Some of the problems are discussed in the following sections.

12.2 BOUNDARY PROBLEM

Boundary problems arise from dichotomy of policy perspectives like moral and immoral, efficient and inefficient as indicated before. Actors in a policy can play different roles from policy formulator to user group or both. If actors play both the roles (formulator and user), the gap of understanding of policy purposes between formulator angle and user angle narrows down, and thus the outcome of policy is bound to be more practicable and user biased. But in practical societies, formulators are few whereas the users are large comprising ordinary mass. Thus, if the same actors occupy both the spaces, the policy outcome could be biased to few and may exclude the interest of a larger mass. Therefore, an alternative way to overcome the problem is to select different actors from both formulator and user groups and integrate them in the policy formulation process.

The basics of the problem (written policy to practice of policy) also includes considering the involvement of actors from the point of policy formulation and implementation. There are people who participate in the "policy coalition" but do not have a common belief. They participate because they have special professional experience to offer. The role of these people is only professional involvement. It may be hard to draw a boundary between their direct participation and participation through the coalition. The boundary problem also comes out in the distinction between "advocates" and "brokers." Policy brokers mediate between coalitions and search for an acceptable solution, e.g., top-level bureaucrats. Sabatier (1987) mentioned that there could be some interest of brokers in the policy, and thus they may sometimes advocate achieving negotiation in a particular direction from which they or their clients can achieve undue benefits. The differences between brokers and advocates thus cannot be defined by an empirical question. Some elements within the actors can be shown as advocating while the other could be brokering. Such behavior does not give a rational stand to classify actors in a particular class (broker or advocate); however, the tendency of actors has been termed as "discursive affinity" for the purpose of this thesis.

In the environmental sustainability paradigm, no strong belief or ideology has yet developed, particularly in developing countries, which can influence the political actors to work for mere environmental sustainability. In concrete political situations, actors make certain political utterances to position themselves in a specific situation, emphasize certain elements, and play down others or avoid certain topics and agree on others. The positioning of political actors is more obvious at certain levels but they are obscure in their practice, particularly in developing countries, which sometimes affects the specific policy. Therefore, it may be difficult to put a boundary on the "discursive affinity" and "discursive positioning" to understand the nature of the influence an actor could make in the policy process of environmental sustainability. In fact, the discursive positioning is more apparent in top-down policy process, whereas discursive affinity develops in the participatory approach to the policy process.

12.3 PROBLEM WITH SOCIAL CONCERN

Social concerns of environmental sustainability may be treated as a broad value change, a shift from the value preferences of goals related to material values (economic and physical) to "postmaterial values" (satisfaction of intellectual needs). These changes were identified by Inglehart's (1990) assumption, which implies:

People are assumed to fulfil their material needs first and only then to aspire to post material goals.

His research presupposes that people value most highly those things that are in short supply (the principle of diminishing marginal utility). The feelings for nature come only when it strikes their health or property or crops. This feeling is rather an individualistic expression of scarcity. Thus, emphasis should be given to structural foundation for explaining social scarcity as an explanation to the problem of evaluation; but in practice, when an evaluation is made, the actions of states get dominance, the interaction between the society and the institutions seldom gets reflected.

12.4 ROLE OF SCIENCE

In the realm of policy, scientific information is commonly used for policy formulation or policy evaluation. Because of some limitations and bias of the application of science, it is likely that only a selective quantity of all known information is utilized for policy. Wynne (1994) pointed out that communication of science is not always independent. Scientific information is reported through intermediate links like newspapers, reviews, and so on where the centralized power is supposed to be no longer a function of the sovereign but is seen as an inherent in all kinds of social practices. Thus, communication of knowledge can be treated as an interactive process instead of one-way process, and therefore, what the receiver or policy actor understands or interprets from the knowledge is important and could have marginal variation on what the investigator tried to communicate. Kummer (1995) reported such biased use of forest statistics in the Philippines. On the other hand, the scientific knowledge comes out of research which is usually funded by and commissioned by the policy-making institutes, and which implies that scientists work within frameworks that are defined by the decision makers. As a result, most researches are only specific. Furthermore, the actual research is often closely monitored by a steering group, which may exert influence on the actual production of scientific knowledge. Also knowledge is often translated, reported in one language but used in another language, and this may bring a break in understanding.

Jasanoff (1990) has drawn a further distinction between proper science and regulatory science to specify the interface in the course of policy evaluation. Clark and Majone (1985) mentioned that distinguishing the science discourse into "numinous" and "civil" legitimacy may facilitate the assessment of reflexivity of science–policy interface in the sustainability controversy. The "numinous" form of legitimacy is derived from "superior authority held to be beyond questioning by those who endure the consequence of power." "Civil legitimacy," on the other hand, is determined by the degree to which scientific claims secure credibility in public debates and are related to the way in which they are played out against other expert opinions or general political or social concerns.

The use of scientific discourses is possible in a face-saving fashion by keeping the real information hidden from the general mass (e.g., by setting lower standard or bypassing expression such as tree health to show large number of healthy trees). For example, statistical information may have a different interpretation to ordinary people if presented from a different level such as national, provincial, or local. If a species is totally eliminated from a locality, e.g., by acid rain, in an expression of national average it may not be reflected. Grainger (1996b) concluded that the inclusion of such biased national forest data in the FAO report of tropical forest assessment of 1990 distorted the estimates of the report. Thus, policy evaluation may cause controversy depending on what forms of scientific knowledge have been used.

12.5 INSTITUTIONAL DIFFICULTY

It is usual that the monitoring responsibility of policy implementation is delegated to a few institutions or organizations rather than individuals. Often, in the case of land use, policy is conducted on the basis of instiualization of sustainability discourse either by self or by some instruction of regulatory institutions. This may extend from identification of the scale of damage done to the forest and determination of threshold level. The identification of damage is largely depends how the problem is defined. For example, the causal gas for the acid rain problem if considered as only SO_2 (sulfur-di-oxide), leaving aside NO_x (nitrogen oxides), NH_3 (ammonia), and O_3 (ozone) complexities, the solution would be one directional to reduce the sulfur gas emission (Hajer, 1995). Similarly, if the policy evaluator starts evaluating with a preoccupied knowledge, for example, deforestation is caused by humans, the other problems like political problems may remain unaddressed or underaddressed. Thus, in some problems, though they happen from a specific source, the blame transfers to the institution or society in general. These are the institutional problem because often institutions fail to integrate all the information into a decision.

12.6 IMPLEMENTATION PROBLEM

The implementation of a policy for achieving the objectives is not always possible to translate in the field—the gap between the ideal and the reality remains. The difference may be called as the "implementation deficit." In relation to this issue, Weale (1992) has drawn a distinction between "policy output," that is the product of the government activity in the form of regulations, laws, inspections, and procedures; and "policy outcomes," that is the material changes which actually takes place as a consequence of the policy outputs. There could be multiple reasons for which the differences may occur, some of which are discussed in the following sections.

12.6.1 Circumstances External to the Implementing Agency
According to Hoggwood and Gunn (1984), there are obstacles to the implementation of policies, which are outside the control of

the administrators. For example, agricultural policies may face intractable problem of implementation due to inclement climate, disease, and/or problems of farm structure. Other constraints may be linked to the vested interests of the pressure groups. For example, land reform policies in many countries faced obstruction from large land owners. Also the problem may occur from outside the political boundary, e.g., policy outcome of land use policy for North Bengal, Bangladesh became different from policy output due to water withdrawal from the Ganga flow by the Farakka Barrage in India.

12.6.2 Inadequacy of Time, Resources, and Programs
In the context of sustainability evaluation, there are a number of constraints originating from policy making and implementation such as lack of institutional resources and access to information; lack of ecological, technological, and administrative knowledge; lack of material or legal resources; and weakness of institutions in relation to vested interests. Effective policy for sustainable development requires the integration of political legitimacy, analytical competence, and administrative capacity to translate policy objective into effective sustainability actions. Even when major physical and political constraints are not present, some policies may fail due to lack of funding or unrealistic expectations that cannot be achieved in a specified period of time. For example, there are many states who are signatory to the Agenda 21 of the Rio protocol but are not able to divert fund toward conservation for sustainable action. Ex-ante evaluation will show such policies very much sustainable but not the ex-post analysis.

12.6.3 Lack of Understanding Between Cause and Effect
Often the policy to be implemented is based on a valid theory of cause and effect but the relationships between the cause and effects are very few. This is more prevalent if the assumptions on causes and effects of policy makers are implicit and unstated. If a policy is taken, not for solving a problem, but for preventive action of an anticipated future problem, it is likely that implementers will be confused and may hesitate to implement the policy unless the purpose is stated clearly. Similarly, if the understanding of the policy makers is inadequate policies may fail. Winter (1996) has drawn some example of policy failure due to lack of understanding.

12.6.4 Minimum Dependency Relationship of Decisions

The relationships between the dependency of decisions and various organizations describe the success of the policy failure. Pressman and Wildavsky (1973) argue that the greater the number of decisions required by different actors at different points in the implementation process, the more likely there is to be a policy failure. Thus multiple-dependency relations among different agencies can cause significant policy blockage. Hoggwood and Gunn (1984) say,

> it is now-a- days relatively rare for implementation of a public programme to involve only a government department on the one hand and the group of affected citizens on the other. Instead there is likely to be an intervening network of local authorities, boards and commissions, voluntary associations, and organised groups.

A contrast between the single ministry and more diffuse environmental policies may be drawn as an example. Though the ministry is usually responsible for planning, the implementation of diffuse environmental policy involved with two or three tires of organizations and diverse community, sometimes the policy presents a considerable dependency relationship problem such as complexity in organizing administrative arrangements.

12.6.5 Lack of Understanding of, and Agreement on, Objectives

Sometimes policies lack clear objectives. The objectives of organizations or programs are often difficult to identify or are vague and evasive. The challenge to pursuing a sustainability evaluation is the different viewpoints, needs, interests, information, and power of the different actors involved in the policy process. For example, policy makers have to balance the political imperative of securing and expanding their power base with a commitment to sustainability because the public's response to policy problems may be driven by their own economic considerations. Even official objectives, where they exist, may not be compatible with one another. The possibility of conflict or confusion remains high when professionals or other groups try to realize their own unofficial goals within a program. This could happen due to personal disagreement with objectives or the official acts under personal interest. Evaluation of those flaws is difficult.

12.6.6 Policy Tasks not Specified in Correct Sequence

Disarray in policy tasks in "policy breaches" among different stages of the policy process—usually known as "policy breach" is a common problem in sustainability evaluation. First, policy breach comes from the difference between decision making and policy implementation. While decision making is a top-down process in which policies are driven by overall goals and political priorities, policy implementation is essentially a bottom-up process involving middle- and low-level managers concerned with organizational needs and day-to-day management requirements and restrictions. The policy breach between the two processes leads to the disintegration between decision making and policy implementation and produces incoherent, ineffective, and inefficient policies. Policy breach may also exist between policy making and policy evaluation. Policy evaluation has not been routinely applied for most policy decisions, and when it is conducted, it is often motivated by procedural requirements or political considerations and thus fails to contribute to continuous policy learning. The third policy breach is between political process and policy process. The policy process is frequently disrupted by changes in the political process. There should be some specified procedures and sequence for the implementation of decisions without which apparently feasible policies may fail. For example, after clear felling an area, a series of actions are required at an appropriate time to replant it such as, procuring seed, growing seedlings, and clearing land. Some of these actions may need to have been undertaken before the actual felling operation started. If the sequences and timings of work are not maintained, the plantation is likely to fail due to lack of rain or excessive cold. Thus, the outcome of plantation failure will be seen but the cause of untimely action will be difficult to identify in policy evaluation.

12.6.7 Lack of Perfect Communication and Coordination

The potential of conflicts among the key components of sustainability poses a challenge to policy evaluation. For example, poverty alleviation policies may result in increased resource use and habitat destruction, thus creating more environmental problems; strict environmental regulation aiming at slowing down environmental degradation may cause the close-down of many small and median enterprises that are main contributors of local economic growth and employment. The segmented outlook of policy making in different sectors further exacerbates these potential conflicts. For example, the agricultural sector may promote agricultural production at the expense of decreased water availability for the industrial sector and for household consumption. Most governments, NGOs, and international agencies do not deal with the social, economic, and environmental needs across different sectors. Policy objectives should be understood by the implementing agencies or by recipients. It is often said that perfect communication and coordination are unattainable particularly in the case of top-down policy processes. The main reason is that information and understanding of policy may get lost when they are communicated from person to person or office to office. Inevitably the changes are reflected through their implementation stages, which is an important characteristic of imperfectly communicated policy. If a change is continued, it is unlikely that the policy actions will be coordinated. Under that circumstance, policy recommendation will be one but the policy outcome could be different due to changed action, thereby the action will remain undetected.

12.6.8 Rare Perfect Compliance of Implementing Body

Sometimes compliance of the members, even within the implementing agencies, may not be perfect for policy implementation. The more radical the policy, the more likely is interagency rivalry. For example, when the policy of a participatory forestry approach was proposed for implementation in Bangladesh, initially traditional forest officers resisted it, and the implementation was very slow and delayed. In practice, the transition from decision to action in the policy process is neither smooth nor obvious (Jenkins, 1978). They may partially succeed and partially fail and there may be unintended side effects which negate the original stated aims.

Although the sustainability evaluation of a policy needs to consider so many factors, most evaluations are done on the basis of few parameters: population, institution, infrastructure, and economics. No doubt, population and infrastructure are important long-term determinants of general resource use trend over decades, but interannual variability of resource use is not tightly linked to such parameters. While the variation in population and institution is gradual, evidences of resource extraction (e.g., deforestation) are irregular, from time to time. Econometric variables such as variation in price structures and local market demands are likely to be better predictors of resource extraction and abandonment rates (Peluso, 1992), thus there are changes in resource use over short-time cycles. This information may be coupled with information of longer-time cycle, e.g., historical information of land use change and policy measures evolved from aforementioned activities and socioeconomic desire for building a diagnostic model of deforestation process. Such models can provide:

- An improved understanding of the factors which control periodical variation in resource extraction and hence their significance on environmental changes.
- An improved understanding of the factors which determine the balance between resource transformation (e.g., new farmland through deforestation and the creation of new farmland by clearing secondary fallow forests).
- An improved understanding of the factors which control the balance between resource extraction and abandonment.
- A method for spatial and temporal extrapolation of the result from the activities ascribed above.

But availability of information and data often limits the historical evaluation of policy to a few factors. Though the evaluation can be extended to local, national, and regional levels, in this thesis the levels will be limited to only national and regional levels. Regional levels will be used as explanatory examples toward supporting national evaluation. The limitation of undertaking studies below national level is that, perhaps, the findings cannot be calibrated against a generalized scale.

CHAPTER 13

Discussion and Recommendation

13.1 DISCUSSION

13.2 RECOMMENDATION

13.3 IMPORTANCE

13.1 DISCUSSION

Sustainability assessment of policy is a discursive process involving social, economical, political, and historical issues associated to resource use, trade, market, and politics. Complexity in assessing sustainability, environmental sustainability in particular, is due to uncompensated environmental externalities involved in policy process (Eckersley, 1996). Many issues in the environment cannot be valued. When such an assessment approach is historic−hermeneutic, unempirical, or comparative, sustainability assessment becomes even more difficult; because, in those cases sustainability assessment of a policy largely includes the qualitative expression on complementarities of objectives delineating commitment, participation, and the lucidity of decision maker, interest groups, and the general public. Thus, sustainability assessment is not straightforward.

Assessment of environmental sustainability of a policy investigates the boundaries between the clean and the polluted, the moral and the immoral, the efficient and the inefficient, the appropriate and the inappropriate accompanying with homogenization (making problem understandable) or heterogenization (operating up in discursive categories) of discourses. To construct a policy discourse and to understand it as the unintended consequences of the interplay of actions is one thing, the other is more interesting to observe how seemingly technical positions of a policy conceal normative commitments, and finally to find out which categories exactly fulfil the sustainability role. That is to say sustainability assessment involves investigating which measures/institutions

Sustainability Assessment. DOI: http://dx.doi.org/10.1016/B978-0-12-407196-4.00013-1

allow the policies to fulfil commitments, how the accomplishment effects could occur, and which course of affairs can be furthered in achieving a sustainable target. Thus, sustainability assessment of policy needs to be oriented to objectives and functional process of policy components.

For the purpose of this study, the functional process of ingredients of a policy can be seen as linkages summarized in Fig. 13.1. The figure is laid out to show how policy evaluation for sustainability needs to deal with objects of different level, different importance, and different stages. There are specific linkages between and among the stages of policy process. Thus, a policy may be evaluated at any stage (e.g., formulation, implementation, monitoring, or outcome) or for any linkage or relationship; however, sustainability assessment should be an integrated outcome overall.

Figure 13.1 hypothetically represents the ingredients of policy functions by a flying butterfly-shape diagram. The policy is shown as the body (contains policy process, which involves the government and the people). Government is represented by the head, because all the regulatory and control functions of a policy are conducted by the government and the people are represented as the tail because they enjoy the benefits as well as participate in the process. Thus, an assessment of the government's role and peoples' participation in the policy process

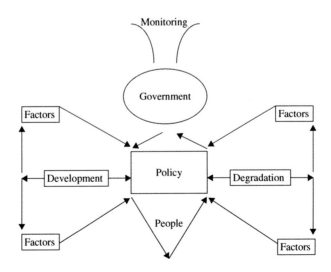

Fig. 13.1 Functional relationship of a typical policy (Hypothetical flying butterfly diagram).

is important for policy evaluation. The monitoring function is shown as the antenna. The diagram also signifies that people are more close to the policy process than the monitoring process. Development as a social process and degradation as an environmental process are represented by the wings, the strength of which will show the sustainability (analogous to flight capability) of a policy. Also they signify that a pair of balanced wings may give optimum result, as is the case in flight better than that of unbalanced stronger wings. Thus, for sustainability, a balanced consideration of both the processes should be involved in the policy process. Both of them are associated with the factors, some are closer to government functions and some of which are closer to peoples' function. The flying mode of butterfly indicates the dynamic nature of policy. Therefore, certain specific criteria may need to be expressed before evaluating the balancing actions of a policy for sustainability.

13.2 RECOMMENDATION

From the discussion, it reveals that assessment of sustainability of resource and environmental policy may be possible in one of the four ways:

1. analyzing the existing situation and future speculation,
2. comparing past to the present after implementation of the policy,
3. comparing among the countries with similar policy, and
4. comparing with a standard situation.

An evaluation on the basis of a single way may not be enough for serving the purposes of assessment of sustainability as outlined in the introduction, particularly for resource policies having long-term effects and impacts which are not easily reversible. Therefore, more than one way of policy evaluation needs to be considered for analyzing the sustainability in relation to resource, environment, and society. Application of different ways of policy evaluation techniques can be successful if conglomerated for an objective evaluation of resource policy for making a clear understanding of relationship between resource environment and sustainability is possible. The relationship between the environment and the resources can be grouped into following four classes:

1. primary source of raw materials,
2. space for waste accumulation and storage,

3. assimilation and regeneration capability for chemically or biologically active wastes, and
4. determinant of environmental quality.

When resource is related to land use, all the four classes of correlates stated above directly relate the resource and the environment and hence it is expected that the evaluation of resource policy should be based on the weights of all types of correlates. In each class of correlates, there may be several factors involved, some of which may be common in all classes of correlates, as long as land use based resource is concerned. As the environmental phenomena of land use involve social processes, a policy is expected to deal with factors that remain active over a certain period of time.

The factors may work independently or interactively or both in the complex of socioenvironmental phenomena. Assessing a consequential evaluation of those interactive factors is quite complex without knowing all the involved factors, their scale and the direction of effects on the environment. However, examples of influences and the cohesive action of such factors that occurred in one society may help to provide a clarification of happenings in another society that may occur due to a similar complexes of factors. Within the same society, the happenings in present socioenvironmental complexes of land use may be explained with the happenings of the past. Depending on such themes, evaluation of land use based resource policy is possible, which may be used as a common directive for planning future actions of sustainable management.

To integrate the discourse of the above theme (temporal) into policy evaluation, the process of story lines followed by some authors (Hajer, 1995) may be pursued for explaining environmental issues, which eventually lead into the history. Such histories bring the past sociopolitical issues producing story lines of environment and ecology. Essentially the observation of Harvey (1993) emphasized that:

> *Ecological arguments are never socially neutral any more than socio-political arguments are ecologically neutral.*

Therefore, a chain of explanation can be given to justify how changes in the environmental phenomena and peoples attitude have happened. Following the examples given by Bryant (1998), resource and environmental degradation can be explained in the way shown in Fig. 13.2.

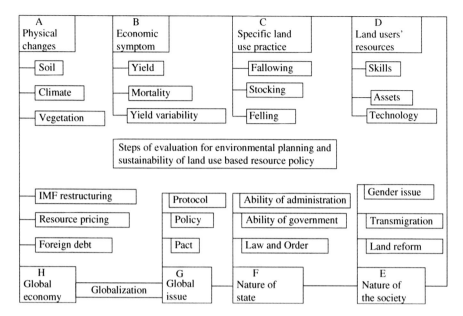

Fig. 13.2 Chain of evaluation for environmental planning and sustainability of forest resource. Adapted from Bryant (1998).

The explanation starts from the physical changes in the resource (Box A) and their associated economic symptoms (Box B). Linking those to specific land use practices (Box C) as well as individual and collective decision-making processes (Boxes D and E), logical build up on wider contextual forces of state policy (Box F), and globalization process (Boxes G and H) may be possible. At each point of the chain of explanation, the ambiguities and complexities associated with understanding and then linking socioenvironmental, political, and economic processes can be emphasized. Thus, addressing environmental sustainability in a policy is a multidimensional task that can potentially induce confusion among the scientists and the policy makers.

In the above discussion, the containment, requirement, and methods of policy evaluation have been presented in a broad spectrum. Often the relevant salient points of those evaluation prospects can be covered in the study of sustainability assessment to avoid complexity and volume of evaluation works. The following steps can be followed for the limited scope of evaluation:

1. Discussing the resource relevant tradition, culture, cohesion, integration, and dissociation of the society.

2. Discussing the present concept, utilization, stocking, state, and distribution of resource.
3. Comparing the past state of a resource to present state outlining policy approaches.
4. Evaluating in the light of examples from the policy approaches of other countries.
5. Evaluating the role of global organization, convention, negotiation.
6. Evaluating peoples response and participation.
7. Evaluating the role of present socioeconomic situation and speculating on the future of sustainability.
8. Discussing the actor influences.
9. Assessing the adequacy of legal and social systems.

Both quantitative and qualitative assessments can be used, wherever their applicability is possible. Validated secondary data, mainly compiled by UN organizations and the statistical bureau of related countries can also be used for quantitative assessment; because, it is often difficult to collect nation-wide information within the limit of the cost and planning. Also for historical information, reliance can be placed on secondary evidences. For qualitative information, evidences of secondary sources can be referred accommodating the practical and social experiences of the researcher as part of the society.

The factors and processes presented in this book show that consideration of all the elements of policy climate in a single policy sustainability assessment is difficult and very complex. Most sustainability assessment considers a few elements relevant to the purpose of policy evaluation. Unlike our proposition in the introduction section, MacRae (1980) has treated policy evaluation as the way to choose the best policy from among a set of alternatives with the aid of reasons and evidences. A consideration of the following elements may identify a best policy from the set of policies:

• importance of problems,
• criteria for choice,
• alternative models and decisions,
• political feasibility.

These are sequences that can be adopted when sustainability paradigm is not in its full force. However, the evaluation of policies still considers the social and economical elements described in earlier sections. Selection of a best policy from a set of policies firstly depends on the scale and importance of the objectives achievable by the policy. If the objectives are not very important for sustainability, then usually cost–benefit is taken into consideration for selection of policy. In addition, criteria or variables on which the evaluation will be made, or a political gain is possible, are considered for selection. It is usually assumed that there are some values behind the formulation of a policy. First, the values need to be given a clear, consistent to general dimension and then the values of the policy can be compared. Not every policy decision involves a value system consistently. Therefore, sometimes multiple value systems need to be used for sustainability evaluation.

Thus, the valuative discourse of policy evaluation may need to be extended from the conceptual to logical definition of values used for their measurement. The specification of a procedure for measuring valuative criteria is one way of defining it clearly. One method of valuation is economic but it depends mostly on the concept of satisfaction and preference. An alternative measurement of sustainability is the measurement of well-being, which is subjective to social indicators. Here, the satisfaction can be measured by questioning individuals. However, the combination of the two valuative criteria entertained by a particular decision maker can also be approached through multiattribute decision analysis. This method concerns the combination of known information and relationships so as to predict the consequences of policies.

Dubnick and Bardes (1983) have suggested that public policy evaluation involves the application of problem-solving techniques to questions concerning the expressed intention of the government and the actions it takes or avoids in attaining those objectives. This broad definition was based on several characteristics of policy evaluation:

1. Policy evaluation is applied.
2. Policy evaluation helps in solving problems.
3. Policy evaluation deals with a variety of questions.
4. Policy evaluation deals with expressions of government intentions or actions taken to achieve policy objectives.

Two other important nonexperimental procedures recommended by MacRae (1980) are: "path analysis" using econometric principles, and "time-series analysis" involving the time period interrupted by the policy intervention. Time-series analysis provides useful information particularly about delayed effects. Policy evaluation is also concerned with the government functions/actions/condition under which they should be taken as implemented. Condition for their efficiency, the implementation ability, and the possible establishment of structure will also facilitate the proper choice and implementation of policies. In relating to sustainability evaluation of policies the following government functions may be considered:

1. direct monetary transactions including tax and subsidies,
2. production and delivery of goods,
3. delivery of service,
4. regulation,
5. monitoring and enforcement,
6. persuasion and socialization, and
7. metapolicy.

These government functions have multiple roles and objectives subsumed within the idea of improving quality of life experienced by different communities (Daniere and Takahashi, 1997). These objectives include the provision of public goods, arbitration of community conflicts, and adoption of policies promoting social welfare. At the same time, government has an ethical obligation to respect the community and cultural values which the society represents and leads. Thus, the improvement of sustainability is facilitated by government actions particularly augmented by the policy purpose. In that case, there is no doubt that a sustainability assessment of the policy may differ for scientific, professional, political, administrative, and personal reasons depending on the purpose for which the policy has been evaluated; however, sustainability assessment for society should not differ much on an individualistic view and should overcome such difficulties.

13.3 IMPORTANCE

From the discussion above, it is understandable that a policy can be evaluated for many reasons. However, the importance of sustainability assessment of a policy has remained greater than other forms of evaluation to avoid several chronic problems of power structure. They are:

- Usually the attention of people and policy makers is attracted toward economic policies than other policies that might not economically significant but have more contribution to sustainability. So sustainability assessment distributes the attention of people and power over all policies.

— Sustainable development acts as a flagship to society whereas policy making is largely driven by social crises. Sustainability assessment may assist in converting policies to prevent crises as well as to continue progress.

— Policy failures lead the frequent changes in political leadership, but the root causes of these failures are associated with sustainability issues that often policy proponents forget to address. Sustainability assessment may prevent such unexpected changes.

— The effects of policies championed by a particular ministry are undermined by strategies employed by another ministry deliberately or otherwise. Elimination of such discrimination would be possible through revealing sustainability potentials of each of the policies and their interrelationship.

— Often in developing countries, policies are formulated and compromised to secure the support of politically powerful groups at the expense of long-term public interests that are underrepresented in the political system; because, sustainability assessment put attention to hidden agenda and emphasize peoples benefit, such bias may be avoided in policy formulation.

The purposes of integration of sustainability assessment objectives with policy evaluation are the following:

— To balance essential aspects of sustainable development in policy making. Different policy options may rank very differently based on differential objectives, and the integration of the objectives ensure that the selected policy action is acceptable with regard to each of the objectives.

— To identify remedial strategies as a part of policy package. Based on all objectives it may not be possible to identify policy options that can produce satisfactory outcomes, but it might be feasible to provide remedial measures to overcome deficiencies in the selected policies.

— To articulate trade-off among different objectives. Given the conflicts among different objectives policy makers may have to choose among tools with unavoidable trade-off. Integration of policy

objectives better inform the decision makers and the public by articulating the trade-off.
— To identify innovative policies by exploring the synergies among different objectives.

Adding more objectives to policy making may not necessarily constrain policy choices. For example, adding the consideration of environmental protection may point to new policy choices which are superior to the original ones because of the synergies among different objectives. The case of sustainability also has synergic integrity, adored and supported by present society to reduce risk and to save the existence of humanity.

SUMMARY

The world is facing a host of environmental and social challenges that are affecting the sustainability of both resource and environment and threatening the existence of human. These include climate transformation, land degradation, air and water pollution, resource depletion, poverty, and hunger. Unsustainable patterns of consumption and production including inefficient use of resources act as significant contributor to these challenges. Policy-makers are increasingly urged to incorporate considerations of sustainable development in their policy decisions. Many governments have developed and implemented projects, programs, and policy measures aiming at improving the impacts of unsustainable production and consumption. However, the majority of measures have been ad hoc, resulting in outcomes that are often not only disconnected from socioenvironmental policies but also in some cases work at odds with them. The central point to these problems is perhaps lack of enough attention paid to sustainability assessment of policy before implementing them.

Sustainability assessment informs managers for taking appropriate policy decision for sustainable development. This is done through evaluating the environmental, social, and economic impacts of policies. The key innovation of sustainability assessment is the emphasis it places on integration of policy objectives with other key social components of sustainable development, such as economic development and poverty reduction. While policy evaluation is primarily concerned with ex-post analysis typically undertaken after the policy change, more attention is increasingly given to ex-ante assessments which are conducted before policy change. The strength of such ex-ante assessments is to provide policy-makers forward looking information allowing them to develop a coherent and integrated set of policy decision. Such ex-ante assessment is important for strategic planning and sustainable policy implementation. The development of sustainability assessment recognizes that pursuing sustainable resource and environmental initiatives without careful considerations to ex-ante development imperatives run the risk of alienating policy agenda from the mainstream decision-makers.

Ex-ante assessment of resource policies is very important for environmental sustainability. Conservative resource use, controlling human attitudes toward resource usage, and positive participation of stakeholders guarantee sustainability. Policies though attempt to formulate allocation and access to resources, they cannot assure sustainability due to dynamic changing of anthropocentric interest and demand. Resources usually claim more than generation time to complete a cycle. Policy actions of today reflect the consequences on resources after a generation when corrections become impossible. Therefore, sustainability assessment, particularly relating the resource and environmental policies is important before implementation as well as on continuous basis along with implementation. Accordingly, incorporation of dynamism in policy actions is also very important to avoid future problems. This study discusses about the elements, factors, and considerations required to maintain dynamism through evaluation of policies and implementing instruments for sustainability. The integration of policies with resource, society, administration, politics, formulation, stakeholders, and players are so vast that dealing with all the aspects have not been possible in a single text. However, a comprehensive overview has been reflected in this publication.

The anthropocentric transformation of resource and environment is in essence of crises of policy implementation. Often very good policies end up in exhaustion of resources and degradation of environment due to nonevaluative undertakings in implementation and perhaps lack of undertaking the dynamics of social system into policy consideration. This study emphasizes on themes that are crucial for analytical problem and also of the integrated understanding of sustainability embedded in policy activity and social domains. Initially, the discussion is focused on the dynamics of policy process, formation, and implementation with their progression of social modernization in general. Influences and imperatives of global trends, trade, treaties, market, and economy on policy processes are also drawn for a wider validity of social modernization. Gradually, the discussion has been developed on elements and factors need to be considered for evaluation of performance of policy implementation for sustainability assessment. This document also explains the coverage, cognition, and criteria of classification to reflect on calculating the indicators associated with resource

and environmental policies. They elaborate on the meaning and interpretation of the concept of social and sustainable situation. It is
expected that the elaboration introduced in this text would be useful
for introducing a new approach for better understating of policy
impacts on sustainability.

REFERENCES

Adger, W.N., Whitby, M.C., 1993. Natural-resource accounting in the land-use sector—theory and practice. Eur. Rev. Agric. Econ. 20 (1), 77–97.

Agarwal, A., 1998. Globalisation, civil society and governance: the challenge for the 21st century, lecture delivered at NORAD's environment day. Norwegian Forum for Environment and Development, 15 December 1998, Oslo (Unpublished).

Ali, M., 2002. Scientific forestry and forest land use in Bangladesh: a discourse analysis of people's attitude. Int. J. Forest. Rev. 4 (3), 214–222.

Ali, M., 1997. Status of environmental legislation in Bangladesh. Chittagong Univ. J. Law 2, 59–78.

Allen, J.C., Barnes, D.F., 1985. The causes of deforestation in developing countries. Ann. Assn. Am. Geogr. 75 (2), 163–184.

Amelung, T., 1991. Tropical Deforestation as an International Economic Problem. Paper presented at the Egonsohmen-foundation Conference, Economic Evolution and Environmental Concerns, 30–31 August, Linz, Austria.

Amelung, T., Diehl, M., 1992. Deforestation of tropical rainforests: economic causes and impact on development, 241. J.C.B. Mohr, Tübingen, Kiel Studies.

Anderson, B.F., 1971. The Psychology Experiment. Brooks-Cole, Belmont, CA.

Anon, 1999. Addressing the underlying causes of deforestation and forest degradation: case Studies. Analysis and Policy Recommendation, Conference report. Available from: <http://www.bionet-us.org/uc-rpt_intro.htm>.

Armostrong, J.S., 1998. Forecasting for environmental decision making. In: Dale, V.H., English, M.R. (Eds.), Tools to Aid Environmental *Decision Making*. Springer-Verlag New York Inc., New York, 296 pp.

Atkinson, G., Dubourg, R., Hamilton, K., Munasinghe, M., Pearce, D., Young, C., 1997. Measuring Sustainable Development. Edward Elgar, London.

Bamberger, M., 1991. The politics of evaluation in developing countries. Eval. Prog. Plan. 14 (4), 325–339.

Barbier, E.B., Burgess, J., Aylward, B., Bishop, J., 1992. Timber Trade, Trade Policies and Environmental Degradation, LEEC discussion paper 92–01, London Environmental Economics Centre, London.

Barrett, S., Fudge, C., 1981. Examining the policy–action relationship. In: Barrett, S., Fudge, S. (Eds.), Policy and Action: Essays on the Implementation of Public Policy. Methuen, London, pp. 3–32.

Bautista, G.M., 1990. The forestry crisis in the Philippines: nature, causes and issues. Dev. Econ. 28 (1), 67–94.

Beck, U., 1992. From industrial society to the risk society: questions of survival, social structure and ecological enlightment. Theory Cult. Soc. 9, 97–123.

Beck, U., 1994. Self-dissolution and self-endangerment of industrial society: what does this mean? In: Beck, U., Lash, S., Giddens, A. (Eds.), Theories of Reflexive Modernization. Polity press, Cambridge, MA.

Berghäll, E., Konvitz, J., 1997. Urbanisation and sustainability. Sustainable Development: In Yakowitz, M (ed.), OECD Policy Approaches for the 21st Century. OECD, Paris, pp. 155–163.

Bergquist, G., Begquist, C., 1998. Post decision assessment. In: Dale, V.H., English, M.R. (Eds.), Tools to Aid Environmental Decision Making. Springer-Verlag New York.

Berstein, P., 1991. Policy domains: organisation, culture, and policy outcomes. Annu. Rev. Sociol. 17, 327–350.

Bhaba, H.K., 1994. The Location of Culture. Routledge, London.

Bilsborrow, R.E., Okoth-Ogendo, H.W.O., 1992. Population-driven changes in land use in developing countries. Ambio 21, 37–45.

Binswanger, H.P., 1989. Brazilian Policies that Encourages Deforestation in the Amazon. Environment Department Working Paper No. 16, Washington, DC. Environment Department the World Bank, Washington D C.

Blaikie, P., 1985. The Political Economy of Soil Erosion in Developing Countries. Longman, London.

Blaikie, P., Brookfield, H., 1987. Defining and debating the problem. In: Blaikie, P., Brookfield, H. (Eds.), Land Degradation and Society. Methuen, London, pp. 1–26.

Blaug, M., 1992. The Methodology of Economics. Cambridge University Press, Cambridge, MA.

Boehmer-Christiansen, S., Skea, J., 1991. Acid Politics: Environmental and Energy Policies in Britain and Germany. Belhaven, London.

Bôjo, J., Mäler, K.G., Unemo, L., 1990. Environment and Development: An Economic Approach. Kluwer Academic Publishers, London.

Bolton, R., 1989. Integrating Economic and Environmental Models: Some Preliminary Considerations, Research paper series, RP 126, Department of Economics, Williams College, Williamstown.

Boserup, E., 1983. The impact of scarcity and plenty on development. J. Interdiscip. Hist. 14 (2), 383–407.

Boyden, S.V., 1987. Western Civilisation in Biological Perspective: Patterns in Bio-History. Oxford University Press, Oxford.

Brown, L., 1995. Nature's limits. In: Brown, L. (Ed.), State of the World 1995. Earthscan, London, pp. 3–20.

Brown, S., Gillespie, A.J.R., Lugo, A.E., 1991. Biomass of tropical forests of south and Southeast Asia. Can. J. Forest. Res. 21, 111–117.

Bryant, R.L., 1994. Shifting the cultivator: the politics of teak regeneration in colonial Burma. Mod. Asian Stud. 28, 225–250.

Bryant, R.L., 1996. Romancing colonial forestry: the discourse of forestry as progress in British Burma. Geogr. J. 162, 169–178.

Bryant, R.L., 1997. The Political Ecology of Forestry in Burma 1824–1994. University of Hawaii Press, Honolulu.

Bryant, R.L., 1998. Power, knowledge and political ecology in the third world: a review. Prog. Phys. Geogr. 22 (1), 79–94.

Bryant, R.L., Rigg, J., Stott, P., 1993. The political ecology of Southeast Asian forests: transdisciplinary discourses. Global Ecol. Biogeogr. Lett.(special issue 3), 101–296.

Buchanan, K., 1973. The white north and the population explosion. Antipode 3, 7–15.

Bunker, S.G., 1984. Extractive economies in the Brazilian Amazon, 1600–1980. Am. J. Sociol. 89, 1017–1064.

Byrd Jr., J., 1980. The humanisation of policy models. In: Nagel, S.S. (Ed.), Improving Policy Analysis. Sage Publications, London, pp. 91–100.

Byron, N., Perez, M.R., 1996. What future for the tropical moist forests 25 years hence? Commonwealth Forest. Rev. 75 (2), 124–129.

Callon, M., Latour, B., 1981. Unscrewing the big leviathan: how actors macro structure reality and how sociologists help them to do so. In: Knorr-Cetina, K., Cicourel, A.V. (Eds.), Advances in Social Theory and Methodology: Toward an Integration of Micro and Macro Sociologies. Routledge and Kegan Paul, Boston, MA, pp. 277–303.

Capistrano, A.D., 1990. Macro-Economic Influences on Tropical Forest Depletion: A Cross Country Analysis, PhD Dissertation. University of Florida, Gainesville, FL.

Capistrano, A.D., Kiker, C.F., 1990. Global Economic Influences on Tropical Closed Broadleaved Forest Depletion, 1967–1985. Food resources Economic Department, University of Florida, Gainesville, FL.

Carney, J.A., 1996. Converting the wetlands, engendering the environment: the intersection of gender with agrarian change in Gambia. In: Peet, R., Watts, M. (Eds.), Liberation Ecologies, Environment, Development, Social Movements. Routledge, London, pp. 165–187.

Castle, E.N., 1982. Agriculture and natural resource adequacy. Am. J. Agric. Econ. 64, 811–820.

Chisolm, M., 1990. The increasing separation of production and consumption. In: Turner, B.L., Clark, W.C., Kates, R.W., et al.,The Earth as Transformed by Human Action. Cambridge University Press, Cambridge, MA.

Clark, W.C., Majone, G., 1985. The critical appraisal of scientific inquiries with policy implications. Science, Technology and Human Values 10 (3), 6–19.

Clarke, J., 1995. Population and the environment: complex relationship. In: Cartledge, B. (Ed.), Population and the Environment: The Linacre Lectures 1993–1994 Oxford University Press, Oxford, pp. 7–31.

Cohen, W.J., 1968. Social indicators: statistics for public policy. Am. Statist. 22 (4), 14–16.

Colchester, M, 1993. Pirates, squatters, and poachers: the political ecology of dispossession of the native people of Sarwak. Global Ecol. Biogeogr. Lett. 3, 158–179.

Culyer, A.J., Lavers, R.J., Williams, A., 1972. Health indicators. In: Shonfield, A., Shaw, S. (Eds.), Social Indicators and Social Policy. Heinemann Educational Books, London, pp. 94–118.

Daniere, A.G., Takahashi, L.M., 1997. Environmental policy in Thailand: values, attitudes, and behaviour among the slum dwellers of Bangkok. Environ. Plan. C Govern. Policy 15, 305–327.

Darden, J., 1975. Population control or a redistribution of wealth? Antipode 7, 50–52.

Dargavel, J., 1992. Incorporating national forests into the new pacific economic order: processes and consequences. In: Dargavel, J., Tucker, R. (Eds.), Changing Pacific Forests - Historical Perspectives on the Economy of the Pacific Basin, Proceedings of a Conference Sponsored by the Forest History Society and IUFRO Forest History Group, Durham, NC, pp. 1–18.

De Bruijn, T.J.N.M., Norberg-Bohm, V., (Hrsq) 2004. Sharing Responsibilities. Voluntary, Collaborative and Information-Based Approaches in Environmental Policy in the US and Europe. MIT Press, Cambridge, MA.

Dee, P.S., 1991. Modelling Steady State Forestry in a Computable General Equilibrium Context. Working Paper Series No. 91/8, National Centre for Development Studies, Canberra.

Devaranjan, S., 1990. Can computable general equilibrium models shed light on the environmental problems of the developing countries? Paper for WIDER conference on The Environment and Emerging Development Issues, September 1990, Helsinki.

Devaranjan, S., Lewis, J., 1991. Structural adjustment and economic reform in Indonesia: model based policies versus rules of thumb. In: Perkins, D., Roemer, M. (Eds.), Reforming Economic Systems in Developing Countries. Harvard Institute for International Development Harvard Kennedy School, Harvard, pp. 159–188.

Dias, A.K., Begg, M., 1994. Environmental policy for sustainable development of natural resources: mechanism for implementation and enforcement. Nat. Resour. Forum 18 (4), 275–286.

Dietz, T., Rosa, E.A., 1994. Rethinking the environmental impacts of population, affluence and technology. Hum. Ecol. Rev. summer/autumn, 1 (excerpt from internet).

Douglass, M., 1988. Purity and Danger—An Analysis of the Concept of Pollution and Taboo. Routledge, London.

Dovers, S.R., 1995. Information, sustainability and policy. Aust. J. Environ. Manage. 2, 142–156.

Dovers, S.R., 1996. Sustainability: demands on policy. J. Public Policy 16 (3), 303–318.

Dubnick, M.J., Bardes, B.A., 1983. Thinking About Public Policy: A Problem Solving Approach. Wiley, New York, NY.

Durning, A.B., 1990. Ending poverty. In: Brown, L.R. (Ed.), State of the World 1990. Unwin Hyman, London, pp. 135–153.

Eckersley, R., 1993. Free market environmentalism: friend or foe? Environ. Polit. 2 (1), 1–19.

Eckersley, R., 1996. Markets, the state and the environment: an overview. In: Eckersley, R. (Ed.), Markets, the State and the Environment—Towards Integration. Macmillan, London, pp. 7–45.

Ehrlich, P.R., Ehrlich, A.H., 1990. The Population Explosion. Simon and Schuster, New York, NY.

Ehrlich, P.R., Holdren, J.P., 1971. Impact of population growth. Science 171, 1212–1217.

Fairweather, G.W., Tornatzky, L.G., 1977. Experimental Methods for Social Policy Research. Pergamon Press, Oxford.

FAO 1979. Economic Analysis of Forestry Projects: Case Studies. FAO Forestry paper 17, SUP-1, Rome.

FAO 1988. Agricultural Policies, Protection and Trade: Selected Working Papers 1985–1987. FAO economic and social development papers 75.

FAO 1993. Forestry Policies of Selected Countries. FAO Forestry Paper 115, Rome.

FAO 1997. State of the World's Forests 1997, FAO Rome. Source: <http://www.fao.org/waicent/faoinfo/forestry/SOFO97SE.htm>.

Flint, E.F., Richards, J.F., 1991. Historical analysis of changes in landuse and carbon stock of vegetation in south and Southeast Asia. Can. J. Forest. Res. 21, 91–110.

Flint, E.F., Richards, J.F., 1994. Trends in carbon content of vegetation in South and Southeast Asia associated with changes in land use. In: Dale, V.H. (Ed.), Effects of Land Use Change on Atmospheric CO_2 Concentration: South and SE Asia as a Case Study. Springer-Verlag, New York, NY, pp. 201–299.

Fox, W., 1996. A critical review of environmental ethics. World Futures 46, 1–21.

Freudenburg, W.R., 1992. Addictive economies: extractive industries and vulnerable localities In a changing world economy. Rural Sociol. 57, 305–332.

Freudenburg, W.R., 1998. Tools for understanding the socioeconomic and political setting for environmental decision making. In: Dale, V.H., English, M.R. (Eds.), Tools to Aid Environmental Decision Making. Springer-Verlag.

Gadgil, M., Guha, R., 1992. This Fissured Land: An Ecological History of India. Routledge, London.

Gaventa, J., 1980. Power and Powerlessness: Quiescence and Rebellion in an Appalachian Valley. University of Illinois Press, Urbana, IL.

Geyer-Allély, E., Eppel, J., 1997. Consumption and production patterns: making the change. Sustainable Development: OECD Policy Approaches for the 21st Century. OECD, Paris, pp. 55–67.

Gillis, M., 1990. Forest Incentive Policies. World Bank Forest policy paper, Washington, DC.

GOB, 1995. Development Perspectives of Forestry Master Plan. Ministry of Environment and Forest, Government of the People's Republic of Bangladesh Dhaka, Bangladesh.

Goudie, A., 1984. The Nature of the Environment. Basil Blackwell, Oxford.

Grainger, A., 1993. Controlling Tropical Deforestation. Earthscan Publications Ltd., London.

Grainger, A., 1996a. The role of policy in controlling deforestation in the Philippines and Thailand. In: Grainger, A., Malayang III, B.S., Francisco, H.A., Domingo, L.J., Mehl, C.B., Tirasawat, P. (Eds.), Controlling Deforestation in the Philippines and Thailand. Report to the Economic and Social Research Council, School of Geography, University of Leeds, (Unpublished), pp. 155–194.

Grainger, A., 1996b. An evaluation of the FAO tropical forest resource assessment—1990. Geogr. J. 162 (1), 73–79.

Grainger, A., 1999. The role of spatial scale in sustainable development. Int. J. Sustainable Dev. World Ecol. 6 (4), 251–264.

Gregory, R., 1998. Identifying environmental values. In: Dale, V.H., English, M.R. (Eds.), Tools to Aid Environmental Decision Making. Springer-Verlag.

Grut, M., Gray, J.A., Egil, N., 1991. Forest Pricing and Concession Policies: Managing the High Forest of West and Central Africa. World Bank Technical Paper NO. 143, Africa Technical Department Series, The World Bank, Washington, DC.

Guha, R., 1989. The Unquiet Woods: Ecological Change and Peasant Resistance in the Himalaya. Oxford University Press, Delhi.

de Guzman, R.P., Padilla, P.L., 1985. Decentralisation, local government institutions and resource mobilisation: the Philippines experience. In: Hye, H.A. (Ed.), Decentralisation, Local Government Institutions and Resource Mobilisation. Bangladesh Academy for Rural Development Comilla, Bangladesh, pp. 138–171.

Hajer, M.A., 1995. The Politics of Environmental Discourse: Ecological Modernization and the Policy Process. Clarendon Press, Oxford.

Harrison, P., 1992. The Third Revolution: Population, Environment and Sustainable World. Penguin, Harmondsworth.

Harrison, P., 1997. The third revolution: population, environment and a sustainable world—executive summary. In: Owen, L.A., Unwin, T. (Eds.), Environmental Management: Readings and Case Studies. Blackwell Publishers, Oxford, pp. 480–484.

Harvey, D., 1993. The nature of environment: the dialectics of social and environmental change. In: Miliband, R., Panitch, L. (Eds.), Real Problems, False Solutions. Merlin Press, London, pp. 1–51.

Hirsch, P., Warren, C., 1998. Introduction: through the environmental looking glass. In: Hirsch, P., Warren, C. (Eds.), The Politics of Environment in Southeast Asia: Resources and Resistance. Routledge, London, pp. 1–28.

Hoggwood, B., Gunn, L.A., 1984. Policy Analysis for the Real World. Oxford University Press, Oxford.

Holdren, J.P., Daily, G.C., Ehrlich, P.R., 1995. The meaning of sustainability: biogeophysical aspects. In: Munasinge, M., Shearer, W. (Eds.), The Meaning of Sustainability. World Bank, Washington, DC.

Holdren, J.P., Ehrlich, P.R., 1974. Human population and the global environment. Am. Sci. 62, 282–292.

Hood, C., 1986. The Tools of Government. Chatham House Publishers, Chatham.

Houghton, R.A., 1994. The worldwide extent of land-use change. BioScience 44 (5), 305–313.

Howlett, M., 1991. Policy instruments, policy styles and policy implementation: national approaches to theories of instrument choice. Policy Stud. J. 19 (2), 1–21.

Hyde, W.F., Newman, D.H., Sedjo, R.A., 1991. Forest Economics and Policy Analysis–An Overview. World Bank Discussion Paper 134.

Hye, H.A., 1985. Introduction. In: Hye, H.A. (Ed.), Decentralisation, Local Government, Institutions and Resource Mobility. Bangladesh Academy for Rural Development Comilla, Bangladesh.

Hype, J.L., 1994. The social implication of soil erosion. Rural Sociol. 9, 364–376.

ICIDI 1980. North-South: A Programme for Survival. Independent Commission on International Development Issues (ICIDI), Brandtland Commission, UN, Pan, London.

ICIDI 1983. North-South: Common Crisis. Independent Commission on International Development Issues (ICIDI), Brandtland Commission, UN, Pan, London.

Inglehart, R., 1990. Culture Shift in Advanced Democracies. Princeton University Press, Princeton, NJ.

Interfuture (Ed.), 1979. Facing the Futures. OECD, Paris.

Jasanoff, S., 1990. The Fifth Branch: Science Advisors as Policy Makers. Harvard University Press, Cambridge, MA.

Jenkins, W.I., 1978. Policy Analysis: A Political and Organisational Perspective. Martin Robertson, London.

Jewitt, S., 1995. Europe's others'? Forestry policy and practices in colonial and post colonial India. Environ. Plan. D Soc. Space 13, 67–90.

John, K., Ian, M., Piera pere., 1991. Landuse Planning and the Control of Development in Spain – with Special Reference to Catalonia, CEPR.

Johnson, T.R., 1999. Community based forest management in the Philippines. J. Forest. 97 (ii), 26–30.

Jokes, S., Leach, M., Green, C., 1995. Gender relations and environmental change. IDS Bull. (special issue 26), 1–95.

Kahn, J.R., McDonald, J.A., 1990. Third World Debt and Tropical Deforestation. Mimeo, Department of Economics, New York University, New York, NY.

Kasperson, J.X., Kasperson, R.E., Turner II, B.L. (Eds.), 1994. Regions at Risk: Comparison on Threatened Environments. United Nations University Press, Tokyo.

Kathirithamby, W.J., 1998. Attitude to natural resources and environment among the upland forest and Sweden communities of Southeast Asia during the nineteenth and early twentieth centuries. In: Grove, R.H., Damodaran, V., Sangwan, S. (Eds.), Nature and the Orient: The Environmental History of South and Southeast Asia. Oxford University Press, Delhi, pp. 918–935.

Khuraibet, A.M., 1990. Achieving Sustainable Development in Kuwait: The Potential Role of EIA at the Project and Policy levels, PhD thesis. University of Aberdeen (Unpublished).

Kolk, A.N.S., 1996. Forests in International Environmental Politics: International Organizations, NGOs and the Brazilian Amazon. International Books, the Netherlands, p. 336.

Kumar, K., 1993. Rapid appraisal methods. Regional and Sectoral Studies. World Bank, Washington, DC.

Kumari, K., 1996. Sustainable forest management: myth or reality, exploring the prospects for Malaysia. Ambio 25 (7), 459–467.

Kummer, D.M., 1992. Deforestation in the Post-War Philippines. Ateneo de Manila University Press, Manila, p. 178.

Kummer, D.M., 1995. The political use of Philippine forestry statistics in the post war period. Crime Law Social Change 22 (2), 163–180.

Kummer, D.M., Turner II, B.L., 1994. The human causes of deforestation in Southeast Asia. BioScience 44 (5), 323–328.

Lebel, L., Steffen, W., 1997. Human Driving Forces of Environmental Change in Southeast Asia and the Implication of Sustainable Development. An Integrated SARCS Study, Executive summary of the Science Plan prepared on behalf of the Southeast Asian Regional Committee for START – Global Change System for Analysis, Research and Training (SARCS).

Leeuw, F.L., 1991. Policy theories, knowledge utilization and evaluation. Knowledge Policy 4 (3), 73–91.

Linder, S.H., Peters, B.G., 1989. Instruments of governments: perceptions and contexts. J. Public Policy 9 (1), 35–38.

Lohmann, L., 1996. Freedom to plant: Indonesia and Thailand in a globalising pulp and paper industry. In: Parnwell, M.J.G., Bryant, R.L. (Eds.), Environmental Changes in Southeast Asia: People, Politics and Sustainable Development. Routledge, London, pp. 23 48.

Lowe, P., Worboys, M., 1978. Ecology and the end of ideology. Antipode 19, 12–21.

MacRae Jr., D., 1980. Policy analysis methods and governmental functions. In: Nagel, S.S. (Ed.), Improving Policy Analysis. Sage Publications, London, pp. 129–151.

Mahar, D., 1989a. Deforestation in Brazil's Amazon region: magnitude, rate and causes. In: Schramm, G., Warford, J. (Eds.), Environmental Management and Economic Development. Johns Hopkins University Press, Baltimore, MD.

Mahar, D., 1989b. Government Policies and Deforestation in Brazil's Amazon Region. The World Bank, Washington, DC.

Marchak, P.M., 1983. Green Gold: The Forest Industries of British Columbia. University of British Columbia Press, Vancouver, BC.

Markusen, A.R., 1987. Profit Cycles, Oligopoly and Regional Development. MIT Press, Cambridge, MA.

Marsden, D., Peter, O., 1990. Evaluating Social Development Projects. Development Guidelines 5, Oxford, Oxfam.

Mather, A.S., 1986. Land Use. Longman, London.

Mather, A.S., 1990. Global Forest Resources. Belhaven, London.

Mather, A.S., 1997. South-North Challenges in Global Forestry. Working Papers No. 145, World Institute for Development Economic Research (WIDER) of the United Nations University (UNU), Finland.

Mather, A.S., Needle, C.L., 2000. The relationship of population and forest trends. Geogr. J. 166 (1), 2–13.

Meadows, D.H., Meadows, D.L., Randers, J., Behrens III, W.W., 1972. The Limits to Growth. Universe Books, New York, NY.

Meadows, D.H., Meadows, D.L., Randers, J., 1992. Beyond the limits: Global Collapse or Sustainable Future. Earthscan, London.

Merkhofer, M.W., 1998. Assessment, refinement and narrowing of options. In: Dale, V.H., English, M.R. (Eds.), Tools to Aid Environmental Decision Making. Springer-Verlag, New York.

Meyer, W.B., Turner II, B.L., 1992. Human population growth and global land use/ cover change. Annu. Rev. Ecol. Syst. 23, 39–61.

Mills, C.W., 1956. The Power Elite. Oxford University Press, New York, NY.

deMontalembert, M.R., 1995. Cross sectoral linkages and the influence of external policies on forest development. Unasylva 46 (3), 25–37, 182.

Muttalib, M.A., 1985. Decentralisation: local self government institutions and resource mobilisation—the Indian experience. In: Hye, H.A. (Ed.), Decentralisation, Local Government Institutions, and Resource Mobilisation. Bangladesh Academy for Rural Development, pp. 172–199.

Muzondo, T.R., Miranda, K.M., Bovenberg, A.L., 1990. Public Policy and the Environment, A Survey of the Literature. IMF Working Paper, WP /90/56. Washington, DC, International Monetary Fund.

Myers, N., 1993. Population, environment and development. Environ. Conserv. 20, 205–216.

Nagel, S.S., 1984. Introduction: policy analysis research, what it is and where it is going. In: Nagel, S.S. (Ed.), Improving Policy Analysis. Sage Publications, London, pp. 7–13.

Nagel, S.S., Neef, M., 1980. What's new about policy analysis research. In: Nagel, S.S. (Ed.), Improving Policy Analysis. Sage Publications, London, pp. 7–13.

NCEDR, 1998. Tools to Aid Environmental Decisions. National Centre for Environmental Decision-making Research (NCEDR), Oak Ridge National Laboratory, TN.

Nord, M., 1994. Natural resources and persistent rural poverty: in search of the nexus. Soc. Nat. Res. 7 (3), 205–220.

Norton, B.C., Noonan, D., 2007. Ecology and valuation: big changes needed. Ecol. Econ. 63, 664–675.

Nunnenkamp, P., 1992. International Financing of Environmental Protection: North South Conflicts on Concepts and Financial Instruments. Kiel Working Papers No. 512, Kiel, Germany, Kiel Institute of World Economics.

OECD, 1984. Future Directions for Environmental Policies in Environment and Economics. OECD, Paris.

OECD, 1985. Environment and Economics—Results of the International Conference on Environment and Economics. OECD, Paris.

Ojima, D.S., Galvin, K.A., Turner II, B.L., 1994. The global impact of land-use change. BioScience 44 (5), 300–304.

Ooi, Jin Bee, 1990. The tropical rain forest: patterns of exploitation and trade. Sing. J. Trop. Geogr. 11, 117–142.

Osleeb, J., Kahn, S., 1998. Integration of geographic information. In: Dale, V.H., English, M.R. (Eds.), Tools to Aid Environmental Decision Making. Springer-Verlag, New York.

Ostrom, E. (1999) Coping with Tragedies of the Commons. Workshop of Political Theory and Policy Analysis, Centre for the Study of Institutions, Population and Environmental Chance, Indiana University, Bloomington, IN.

Palumbo, D.J., Hallett, M.A., 1993. Conflict versus consensus models in policy evaluation and implementation. Eval. Prog. Plan. 16 (1), 11–23.

Panayotou, T., Ashton, P., 1992. Not by Timber Alone: Economics and Ecology for Sustaining Tropical Forest. Island Press, Washington, DC.

Pearce, D.W., 1990. An Economic Approach to Saving the Tropical Forests. LEEC paper 90–06, London Environmental Economic Centre, London.

Pearce, D.W., Barbier, E.B., Markandya, A., 1990. Sustainable Development: Economics and Environment in the Third World. Edward Elgar and Earthscan, London.

Peet, R., Watts, M., 1996. Liberation ecology: development, sustainability and environment in an age of market triumphalism. In: Peet, R., Watts, M. (Eds.), Liberation Ecologies: Environment, Development and Social Movements. Routledge, London, pp. 1–45.

Peluso, N.L., 1992. Rich Forests, Poor People: Resource Control and Resistance in Java. University of California Press, Berkeley, CA.

Peluso, N.L., Turner, M., Fortmann, L., 1994. Introducing Community Forestry: Annotated Listing of Topics and Readings. FAO, Rome.

Pillai, V.K., 1996. Air pollution in developing and developed nations: a pooled cross-sectional time series regression analysis. Int. Plan. Stud. 1 (1), 35–47.

Porter, G., Ganapin, D., 1988. Resources, Population and the Philippines' Future. World Resources Institute, Washington, DC.

Pressman, J., Wildavsky, A., 1973. Implementation. University of California Press, Berkely, CA.

Rausser, G.C., 1992. Political economic markets: PERTs and PESTs in food and agriculture. Am. J. Agric. Econ. 64 (5), 821–833.

Redclift, M., 1991. The multiple dimensions of sustainable development. Geography 76, 36–42.

Rees, J., 1990. Natural Resources: Allocation, Economics and Policy, second ed. Routledge, London.

Reid, A., 1998. Humans and factors in pre-colonial Southeast Asia. In: Grove, R.H., Damodaran, V., Sangwan, S. (Eds.), Nature and the Orient: The Environmental History of South and Southeast Asia. Oxford University Press, Delhi, pp. 106–126.

Reid, W.V., Miller, K.R., 1989. Keeping Options Alive: The Scientific Basis for Conserving Biodiversity. World Resources Institute, Washington D.C.

Reis, E.J., Marguilis, S., 1991. Options for slowing Amazon jungle clearing. Paper presented in a conference on Economic Policy Responses to Global Warming, Rome.

Repetto, R., 1988. The Forest for the Trees? Government Policies and the Misuse of Forest Resources. World Resource Institute, Washington, DC.

Repetto, R, 1990. Deforestation in the tropics. Sci. Am. 262 (4), 36–45.

Rose, A.M., 1967. The Power Structure: The Political Process in American Society. Oxford University Press, New York, NY.

Sabatier, P.A., 1987. Knowledge, policy oriented learning, and policy change. Knowl. Creat. Diff. Util. 8 (4), 649–662.

Saldanha, I.M., 1998. Colonial forest regulations and collective resistance, nineteenth century Thana district. In: Grove, R.H., Damodaran, V., Sangwan, S. (Eds.), Nature and the Orient: The Environmental History of South and Southeast Asia. Oxford University Press, Delhi, pp. 708–733.

Schneider, R., McKenna, J., Dejou, C., Butler, J., Barrows, J., 1990. Brazil: An economic Analysis of Environmental Problems in the Amazon. World Bank, Washington, DC.

Stebbing, E.P., 1921. The Forests of India, vol. 1. John Lane the Bodley Head Limited, London.

Stern, P., Young, O., Druckman, D., 1992. Global Environmental Change: Understanding the Human Dimensions. National Academy Press, Washington, DC.

Stiles, D., 1994. Tribals and trade: a strategy for cultural and ecological survival. Ambio 23 (2), 106–111.

Summers, L., 1992. Summary on sustainable growth. The Economist, 30 May 1992, 91.

Samuel-Johnson, K., Esty, D.C., 2001. Environmental Sustainability Index. World Economic Forum, Davos, Switzerland.

Strand, J., Toman, M., 2010. Green stimulus. Economic Recovery and Long-Term Sustainable Development. The World Bank, Policy Research Working Paper 5163.

Theodoulou, S.Z., Kofinis, C., 2004. Understanding American Public Policy Making. Wadsworth/Thomson Learning, Boston, MA, USA.

Thiele, R., Wiebelt, M., 1993a. Modelling Deforestation in a Computable General Equilibrium Model. Kiel Working Paper No. 555, Kiel Institute of World Economics, Kiel, Germany.

Thiele, R., Wiebelt, M, 1993b. National and international policies for tropical rainforest conservation—a quantitative analysis for Cameroon. Environ. Res. Econ. 3, 501–531.

Tobey, J.A., 1989. Economic development and environmental management in the third world. Habitat Int. 13 (4), 125–135.

Turner II, B.L., Clark, W.C., Kates, R.W., Richards, J.T., Mathews, J.T., Meyer, W.B., 1990. The Earth as Transformed by Human Action. Cambridge University Press, New York, NY.

Turner II, B.L., Moss, R.H., Skole, D., 1993. Relating Land Use and Global Land Cover Change: A Proposal for an IGBP HDP Core Project. International Geosphere Biosphere Programme, Stockholm, Sweden.

UNDP, 1993. Human Development Report. Oxford University Press, New York, NY.

Uphoff, N., 1985. Local institutions and decentralisation for development. In: Hye, H.A. (Ed.), Decentralisation, Local Government Institutions and Resource Mobilisation. Bangladesh Academy of Rural Development Comilla, Bangladesh, pp. 43–78.

Valadez, J., Bamberger, M., 1997. Monitoring and evaluating social programs in developing countries: a handbook for policy makers, managers, and researchers. EDI Development Studies. Economic Development Institute of the World Bank, Washington, DC.

Wackernagel, M., Rees, W., 1996. Our Ecological Impact: Reducing Human Impact on the Earth. The Society Publishers, British Colombia.

Waggener, T.R., Lane, C., 1997. Asia-Pacific Forestry Sector Outlook Study—Working Paper Series: Asia Pacific Forestry Towards 2010. Working Paper No. APFSOS/WP/02, FAO, Rome.

Walsh, S.J., Rindfuss, R.R., Entwise, B., Chamratrithirong, A., 1999. Population–Environment Interaction in NE Thailand: An Overview of an Ongoing Research. LUCC Newsletter No. 4.

WCED, 1987. Our common future. Report of the Brundtland's Commission for World Commission on Environment and Development. Oxford University Press, Oxford, UK.

Weale, A, 1992. The New Politics of Pollution. Manchester University Press, Manchester.

Williams, M., 1994. Forests and tree cover. In: Meyer, W.B., Turner II, B.L. (Eds.), Changes in Land Use and Land Cover: A Global Perspective. Cambridge University Press, Cambridge, UK, pp. 97–124.

Wilson, G.A., Bryant, R.L., 1997. Environmental Management: New Directions for the Twenty-First Century. UCLA Press Ltd., London.

Winter, M., 1996. Rural Politics: Policies for Agriculture, Forestry and the Environment. Routledge, London, p. 341.

WRI, 1996. World resources 1996–1997. A Joint Publication by WRI, UNEP, UNDP and WB. Oxford University Press, London.

Wynne, B., 1994. Scientific knowledge and the global environment. In: Redclift, M.R., Benton, T. (Eds.), Social Theory and the Global Environment. Rotledge, London, pp. 169–189.